Government
Procurement
and Operations

Government Procurement and Operations

Ivan J. Tether

Environmental Law Institute State and Local Energy
Conservation Project

Ballinger Publishing Company • Cambridge, Massachusetts
A Subsidiary of J.B. Lippincott Company

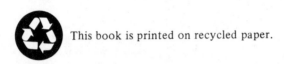 This book is printed on recycled paper.

This book was prepared with the support of NSF Grant APR 7504814. However, any opinions, findings, conclusions, or recommendations herein are those of the author and do not necessarily reflect the views of NSF.

International Standard Book Number: 0-88410-057-X

Library of Congress Catalog Card Number: 77-333

Printed in the United States of America

Library of Congress Cataloging in Publication Data

Tether, Ivan J
  Government procurement and operations.

  (Environmental Law Institute energy conservation series)
  Includes bibliographical references. 1. Public contracts—United
States. 2. Energy conservation—Law and legislation—United States.
I. Title. II. Series: Environmental Law Institute. Environmental Law
Institute energy conservation series.
KF849.T47   346'.73'023   77-333
ISBN 0-88410-057-X

Contents

List of Figures

List of Tables

Preface

For nearly two years, the Environmental Law Institute has engaged in an extensive study of state and local energy conservation. The Institute's Energy Conservation Project has examined ways in which laws and political and social structures influence the consumption and conservation of energy resources. The project sought first to identify these relationships and subsequently to select strategies that showed high potential for conservation. Finally, project members devised original strategies, backed up in part by suggested legislative approaches, for overcoming existing legal and institutional barriers and for encouraging major conservation efforts.

Ivan Tether's book is one of a series of eight that represents a major product of the Energy Conservation Project. The topics of the books have been shaped to fit the energy information needs of government departments in developing effective conservation programs in their respective areas. The books are also intended to assist citizen leaders, legislators, and other interested persons as they work to enhance conservation efforts. This book is intended for use by government purchasing departments and by agencies that are responsible for managing government property. Recent energy programs and proposals have tended to overlook government consumption—out of either myopia or a belief that the "housekeeping" arena is not a significant one. This is an unfortunate oversight. Not only are there manifold opportunities for conservation in governments' own operations, but recent political events have demonstrated that even as government must be by laws, not men, it must also be by example, not rhetoric. We believe that this book will contribute substantially toward unlocking the potential for conservation within government.

> **Grant P. Thompson**
> Institute Fellow
> Principal Investigator
> Energy Conservation Project

Acknowledgments

Grateful acknowledgment is made to my colleagues, authors of the other works of this series. I am honored to have worked with them. Special thanks goes to Grant P. Thompson, a source of not infrequent support and inspiration, and to Jay Russell who shared an office with me—no mean feat! Others who worked with us and deserve special gratitude are Barbara L. Shaw, Stephen O. Anderson, Gail Boyer Hayes, Patricia Sagurton, Janet Coffin, Stephanie Nagata, Tina Luebke, Kathleen Courrier, and Janet Pierson.

Two individuals, Jocelyn Karp and Christine Lipaj, made perhaps the most tangible contributions to my work. Research assistants from Georgetown University Law Center, they drafted a significant portion of the latter sections. Ms. Karp helped me to develop my ideas in early stages of writing. Without the prolific and excellent assistance of Ms. Lipaj, who provided substantial input during the final months, I would never have met the deadline for completion.

Multiple thanks go to Nancy L. Southard, Esq.—my wife. Not only did she provide me with uninterrupted support and inspiration, but she also gave ongoing professional criticism as my work developed.

Our national advisory panel, assembled to review preliminary drafts, provided invaluable "nuts and bolts" criticism. I am deeply indebted to panel members who reviewed my work: Ted Fody, director of the experimental technology division of the Federal Supply Service; Sue Guenther, director of the energy project of the National Association of Counties; John Hittinger, director of the division of purchasing of the state of Florida; and C. Richard Boehlert, deputy associate general counsel of the U.S. Environmental Protection Agency and member of the American Bar Association Coordinating Committee on a Model Procurement Code.

Grateful acknowledgment is also made to Alan Beres, Federal Supply Service (FSS); Joseph G. Berke, National Bureau of Standards; Bill Christiansen, lieutenant governor of Montana; Eldon Crowell, American Bar Association Coordinating Committee on a Model Procurement Code; Larry Frisbee, FSS; Robert H. Fuller, Ohio State University; L. Donald Genova, FSS; Millard Hulse, FSS; Fred Knight, International City Management Association; Mike Malloy, Montana legislature; Joe McElwee, FSS; Frank Miles, Association of Home Appliance Manufacturers; Jon Mills, Center for Governmental Responsibility; Craig Othmer, Federal Energy Administration; and F. Trowbridge vom Baur, American Bar Association Coordinating Committee on a Model Procurement Code.

Government
Procurement
and Operations

Introduction

This book is intended for use by state and local officials who are involved in "keeping the government's house," and by any other individuals who are interested in seeing that this function is carried out so as to promote energy conservation. The book examines legal and institutional aspects of conserving energy through government procurement and government operations. This examination includes analysis of the use of procurement and operations as tools to effect energy conservation in parts of society external to government. The book's format displays a problem-solving orientation. Discussion is organized primarily under strategies —strategies for overcoming various legal, institutional, and political roadblocks to conservation and for affirmatively implementing conservation programs. These strategies are often presented with alternatives, to provide readers with choice rather than dogma.

It is hoped that individuals in many fields other than law will read and benefit from this book. Accordingly, while analysis is often from a legal standpoint, an attempt has been made to explain specialized concepts in ordinary language. Those who are experts in the field of government purchasing may want to skip some of the explanatory material, and those who have no familiarity with this area may benefit from the background readings mentioned in footnotes, but both groups will find that this book has been written with their needs in mind.

ENERGY CONSERVATION

Dozens of "energy" books written since the winter of 1973—1974 begin with a chronicle of the "energy crisis" of that period, and then proceed to suggest divers measures that should be taken to see that the "crisis" does not recur. More recently, however, the literature has matured to a recognition that "crisis" is not an appropriate framework for dealing with our chronic energy shortage. Oil imports have increased from 29 percent of national use in 1973—1974 to about 40 percent of national use in the fall of 1976—and are expected to surge upward in the near future. There is a long term, and perhaps permanent, energy shortage that threatens individual convenience, national security, and worldwide economic well-being.

There are two major approaches to shortage: (1) get more, or (2) make better use of what there is. An economist might call the first approach a supply orientation and the second a demand orientation. In economics parlance, this book expresses a demand orientation—it presents strategies for reducing demand for energy by increasing the efficiency of current uses of energy or, occasionally, by eliminating certain existing uses.

There are forcefully asserted arguments that favor either a supply or a demand approach to the virtual exclusion of the other. Belief in the importance of the strategies in this book does not require adherence to either doctrine. Supply and demand—or conservation—approaches must both be employed to deal with energy shortage. For present purposes, the following assumption will suffice: Energy conservation has a highly significant role to play in mitigating energy shortage, a role that is at least as important as efforts to increase the supply of depletable energy. This assumption is admirably buttressed by Denis Hayes of Worldwatch Institute in his paper, "Energy: The Case for Conservation" [1].

The actual definition of energy conservation is also subject to various interpretations. Definitions include: (1) designing and producing more efficient energy-consuming machines; (2) using energy-consuming machines so that they consume less energy (e.g., using them for shorter periods or at reduced output); and (3) not using certain machines at all. Controversy arises between those who feel that conservation should be limited to increased hardware efficiency and those who feel that conservation requires significant changes in lifestyle [2]. Again, this book demands no clear delineation between these viewpoints. Emphasis here is upon increased efficiency of both machines and operations.

GOVERNMENT PROCUREMENT AND OPERATIONS

State and local governments trying to conserve energy in their respective jurisdictions and to contribute to national energy conservation have far greater control over what they buy, how they use it, and how they dispose of it than over the way that the rest of society carries out these same functions. Within the narrow field referred to earlier as "keeping the government's house," the scope of procurement power is very broad. In the words of Mr. Justice Black, writing for the United States Supreme Court: "Like private individuals and businesses, the Government enjoys the unrestricted power to produce its own supplies, to determine those with whom it will deal, and to fix the terms and conditions upon which it will make needed purchases" [3]. Using this broad power, governments can reduce energy consumption, both their own and that of private entities, in many ways. Strategies for procuring machines and buildings for government use, and for using the procurement power to affect energy use by other sectors of society are presented in Chapters 1 through 5. Strategies for managing and modifying government property so that it uses less energy are found in Chapters 6 and 7. Strategies for conserving energy in transferring and otherwise disposing of government property that is no longer needed are discussed in Chapter 8. The opening chapter of each book segment provides further description of that segment's substance; further description here would be superfluous.

NOTES TO INTRODUCTION

1. Denis Hayes, "Energy: The Case for Conservation," Worldwatch Paper 4 (Washington, D.C.: Worldwatch Institute, 1976). This paper is available from Worldwatch Institute, 1776 Massachusetts Avenue, N.W., Washington, D.C. 20036. Single copy price: $2.00.

2. An interesting study that suggests that reduced energy consumption need not be detrimental to the quality of life was recently performed at the University of California at Berkeley. *See* Lee Schipper and A.J. Lichtenberg, *Efficient Energy Use and Well-Being, The Swedish Example* (Berkeley, California: University of California, 1976). Please address correspondence to the authors at: Room 100, Building T–4, University of California, Berkeley, California 94720.

3. *Perkins* v. *Lukens Steel Co.*, 310 U.S. 113, 60 S. Ct. 869, 84 L.Ed. 1108 (1940).

Chapter 1

Purchasing Strategies for Reducing the Direct Consumption of Energy

INTRODUCTION

Purchasing by any group or individual affects energy consumption. But so great is the overall volume of government purchasing that it may be considered an implement of social change.

Strategies for conserving energy within the context of government procurement, the subject of Chapters 1 through 5, have not as yet been widely implemented [1]. Some strategies presented here, such as life cycle costing, have been successfully put into effect by several states. Others, such as the value incentive clause, have been "borrowed" from federal purchasing procedures. Still others, such as percentage price differentials, are purchasing procedures that have been newly adapted to energy conservation purposes.

All state and local procurement strategies have been given new significance by the Energy Policy and Conservation Act of 1975 [2]. Subchapter III.B of this act provides for federal funding of state energy conservation plans and requires that, to be eligible for funding, such plans must include "mandatory standards and policies relating to energy efficiency to govern the procurement practices of such State and its political subdivisions" [3].

Of no small significance to energy conservation is the American Bar Association (ABA) effort to prepare a Model Procurement Code. The ABA Section of Local Government Law and the ABA Section of Public Contract Law are currently engaged in a joint effort to formulate this code. Preliminary Working Paper No. 1, released during the summer of 1976, seems to provide a sound base for the implementation of energy conservation strategies, and the ABA has received the unanimous support of the executive committee of the National Association of State Purchasing Officials and the board of directors of the National Institute of Governmental Purchasing. Purchasing officials and other interested persons are urged to follow the development of the code [4].

The author has attempted to make this book comprehensible to a wide range of readers who may be interested in it for a variety of reasons. It is intended for use by lawyers, but also by those with no legal training. It is also intended for use by individuals who have no prior knowledge of purchasing. Accordingly, the author has defined terms and concepts throughout the book, but has also tried to avoid the level of detail that might cause the seasoned professional to slump forward at his or her desk, using the book only as a pillow. For those unfamiliar with the purchasing process, or who wish further insight into the perspective taken in preparing this work, a brief sketch of the purchasing process is provided in the appendix to this chapter.

As mentioned above, the first five chapters of this book discuss purchasing. This chapter presents strategies to increase the energy efficiency of government operations, primarily through procurement of more efficient appliances, vehicles, and buildings. These strategies will reduce a government's direct energy consumption and will thereby save money. In many cases, the money saved in utility bills will be sufficient to pay back the cost of implementing the strategy. Consequently, these approaches to conservation are likely to enjoy a high level of political acceptability.

Chapter 2 discusses indirect energy consumption and presents several strategies that the government can use to reduce this almost hidden energy cost. Energy required to produce, transport, and dispose of items used by governments, this chapter stresses, is really consumed by the governments themselves. By reducing this energy use, a government reduces its indirect energy consumption and saves energy dollars. Further, governments can exercise their considerable control over the practices of businesses that provide services under government contract—particularly those services performed at government facilities—in ways that encourage energy efficiency. Examples discussed are trash collection (when performed under private contract rather than by government employees) and the sale of beverages on government property. Since trash collection is performed on behalf of the government, the public views it as a government function, and this visibility makes it a particularly important candidate for responsible energy use.

Chapter 3 covers the extension of the use of contract leverage to control of contractors that supply goods or construct buildings. This strategy assumes that if government contracts are desirable, then contractors will take certain energy-conserving measures to obtain them or to become eligible to obtain them. The precedent for debarment of contractors whose operations do not meet certain energy conservation standards from bidding for government contracts was set in the federal Clean Air and Water Acts, in 1970 and 1972 respectively [5]. In addition to threatening debarment, governments can include in contracts clauses that require adoption of energy conservation practices in performance of the contract.

Chapter 4 discusses two topics: (1) the possible use of energy impact statements to forecast the energy use associated with various public works; and (2) one aspect of energy impact—the location of government buildings and other public works, particularly as relates to transportation of facility users.

Chapter 5 examines cooperative purchasing—a strategy to help purchasing agencies find time and money for other strategies.

To return to Chapter 1, its strategies derive their political acceptability from their cost-saving aspects. While purchasing strategies presented in subsequent chapters may find their cost justification in less tangible benefits, such as encouraging private citizens to conserve energy, the strategies in this chapter can usually be fully justified on the basis of saving money. Furthermore, they often set an example for the private sector, stimulate the development of energy-efficient technology, and curtail the spread of environmental degradation, which often accompanies energy production and consumption.

ENERGY EFFICIENCY STANDARDS

Problem Focus

For many reasons, including higher acquisition prices of energy-efficient items, lack of awareness, and bureaucratic tradition, many state and local purchasing authorities continue to buy goods and buildings that consume unnecessarily large amounts of energy. The aggregate of governmental energy consumption is thus much greater than required by the present level of operations. Reduction of this energy consumption requires a means of identifying the energy efficiency of prospective purchases and a means of selecting items that are relatively efficient.

Strategy

Energy efficiency standards are simple, item-by-item requirements of minimal energy efficiency. Adopted by statute and often amplified by administrative regulations, these standards would prohibit the procurement of energy-consuming products with less than a prescribed efficiency [6].

The mathematical simplicity of this strategy is deceptive, but its success depends on the care with which standards are set. It is easy to pull a number from the air and assign it the status of "energy efficiency standard" for particular items; it is difficult to gauge in advance the level of efficiency that will provide maximum energy savings over the life of an item, with a minimal increase in that item's acquisition cost. This problem is exacerbated by inaccurate perceptions of future market conditions, which may result in:

1. Setting the standard far too low, so that the opportunity is forfeited to acquire energy-efficient items available at a slightly higher cost; the small saving in acquisition price is gobbled up by higher energy expenditures.
2. Setting the standard too high, so that it surpasses what is practicable for a particular product; the item is therefore either unobtainable or very expensive.
3. Setting the standard so high that only one supplier can satisfy the requirement; the absence of competition either violates the purchasing statute or makes the price unnecessarily high.

Even a thoughtfully prescribed standard is unlikely to perform well in every procurement situation. The availability of energy-efficient items, and the price of those items, may fluctuate markedly over short periods. Furthermore, unless constantly revised, these standards will not keep pace with technological advances. Since they are often considered "minimum" standards when set, but "maximum" standards when put into practice, numeric standards have built-in obsolescence.

There are several ways to minimize the problems raised by energy-efficiency standards: (1) by preceding standard setting with thorough analysis; and by frequently updating standards; or alternatively (2) by tying the standards to some external indicator of mechanical practicability. As will be seen by reference to the appendix, many bills specify standards to be met by certain items [7]. This approach is the *least* effective in terms of keeping standards current with new advances in technology.

Administrative regulations should serve as the vehicles for frequently adjusting energy efficiency standards to fluctuating procurement markets. Statutes function better as enabling legislation than as means of prescribing specific efficiency levels. Standard-setting responsibility should, therefore, be vested in the director of the department of general services or other similar agency [8]. While legislators may be tempted to set minimum standards to guide officials vested with standard-setting responsibility, this is probably unwise. As suggested earlier, "minimum" standards have a way of becoming "maximum" standards. It is better if the analysis that precedes standard setting is unprejudiced by legislative standards. If necessary, legislative oversight can best be provided in periodic hearings with testimony by agency officials.

Before setting energy efficiency standards for various items, the responsible official should analyze (1) the energy efficiency of items presently owned by the government; and (2) the state of the art, both generally and with particular reference to the cost of various levels of energy efficiency. This knowledge makes it possible to estimate the potential energy savings of various efficiency requirements, as well as the resultant increase in acquisition costs. When setting efficiency standards, officials will encounter some specialized needs for relatively inefficient items, such as heavy, high-powered vehicles for certain police operations. (Note that relatively small, efficient vehicles can be modified to serve most police needs.) To allow such purchases, the energy efficiency standard may be applied to the *average* energy efficiency of all the items procured in a particular category rather than to every item. Alternatively, specific exemptions may be provided for specialized needs. When energy efficiency standards are tied to an external indicator, their effectiveness depends, not surprisingly, on the validity of that indicator. Use of a valid, frequently updated indicator reduces administrative costs and eliminates the possibility that a standard-setting official may perform ineffectually. Such an external indicator would not, of course, be sensitive to periodic and regional market fluctuations.

The Author's Suggested Legislative Approach Number Two [9] uses this approach by tying auto efficiency (fuel efficiency) to a percentile ranking of motor vehicles according to fuel efficiency by the U.S. Environmental Protection Agency.

The Energy Policy and Conservation Act uses this general approach by requiring that

all passenger automobiles acquired by all executive agencies in each fiscal year which begins after such date of enactment achieve a fleet average fuel economy for such year not less than—

(1) 18 miles per gallon, or

(2) *the average fuel economy standard applicable under section 502(a) for the model year which includes January 1 of such fiscal year, whichever is greater* [10].

There are other possible approaches, such as referring to industry standards or ratings. Even when energy efficiency standards are tied to external indicators, they should be frequently reevaluated to see if they are requiring optimal efficiency.

PERCENTAGE PRICE DIFFERENTIALS

Problem Focus

Most government procurement of energy-consuming commodities involves competitive bidding. Contracts are nearly always awarded to the bidder who offers the lowest acquisition price, with little or no consideration of energy efficiency. Since the least expensive commodities are often the least efficient, money saved at the time of acquisition is usually insufficient to justify increased operating costs. A contract-awarding mechanism is needed that will make trade-offs between acquisition price and energy efficiency.

Strategy

Percentage price differentials are advantages granted to higher priced bids because those bids have a special, desirable quality [11]. Advantage is granted according to a set percentage, which may vary with the type of characteristic involved or the degree of its desirability to the government.

Presently, the main use of price differentials by state and local governments is to favor local or in-state products or bidders. A recent study by the Council of State Governments reports that eleven states have statutes providing percentage preferences to in-state bidders [12]. This practice is rationalized as a boost to local economies.

Applying percentage price differentials to favor energy efficiency is more complicated than applying them to in-state products or bidders because energy efficiency is a relative concept. It is necessary to set up a sliding scale of the percentage price differentials that will be granted for specified increases in efficiency. One possible scheme for implementing this strategy is set out in the suggested legislation in the appendix.

Table 1–1 should help dispel confusion; it also compares the advantages of percentage price differentials with the use of numeric energy efficiency standards. The hypothetical situation in Table 1–1 involves the evaluation of five bids for the supply of air conditioners with 10,000 BTUs cooling power. Evaluation is on the basis of unit price and energy efficiency. Three types of evaluation are demonstrated: (1) least cost, which chooses the lowest priced bid without further consideration; (2) least cost with a minimum energy efficiency

Table 1–1.   Three Ways to Evaluate Bids to Supply Air Conditioners: Least Cost, Least Cost With Minimum EER, and Least Cost Modified by Percentage Price Differentials

First: Least Cost (the traditional me 1od).

Air Conditioner	Acquisition Cost ($)	EER (BTU/Watt)
A	200	(7)
B	220	(7)
C	280	(9)
D	300	(11)
E	340	(12)

Result: Unit A, with low acquisition price of $200, wins the contract. Note that EER, the energy efficiency factor, is extraneous to the evaluation, and is thus enclosed in parentheses.

Second: Least Cost with Minimum EER (assume a minimum EER of 8 has been set by administrative regulation).

Air Conditioner	EER (BTU/Watt)	Acquisition Cost
A	7	200
B	7	220
C	9	280
D	11	300
E	12	340

Result: C, D, and E have acceptable EERs; C, with lowest acquisition cost of the three, wins the contract.

Third: Least Cost Modified by Percentage Price Differentials (PPDs).

Sample Scale for Assignment of Percentage Price Differentials for Given Increases of Energy Efficiency (as compared to that of lowest priced bid):

Energy Efficiency Increase (%):	1	5	10	15	20	25	30	35	40	45	50+
Percentage Price Differential (%):	1.5	3	5	7	10	13	20	23	27	31	35

(For example, a 1 percent EER increase over that of the low bid yields a 1.5 percent price differential; a 15 percent EER increase yields a 7 percent price differential, etc. Note that the ratio is arbitrary and solely for illustration.)

Air Con-ditioner	EER (BTU/Watt)	Energy Efficiency Increase (%)	Acquisition Cost ($)	Cost Increase ($)	PPD (%)	PPD Cost ($)
A	7	—	200	—	—	200
B	7	0	220	20	0	220
C	9	28.6	280	80	13	243.6
D	11	57.1	300	100	35	195.0
E	12	71.4	340	140	35	221

Result: D wins because it earns a $105 price differential (35 percent of $300) for its 57.1 percent energy efficiency increase.

ratio of 8; and (3) least cost as modified by percentage price differentials. The ratios of price percentages to various increments of energy efficiency improvements are arbitrarily set in Table 1–1. Note that a ceiling of 35 percent has been placed on preferences so that there is no *increase* in reward for improving energy efficiency beyond 150 percent of that of the lowest cost air conditioner. This ceiling may fail to stimulate innovation beyond a certain level, but it may make the strategy politically acceptable by setting an upper limit on its cost. (For further discussion of setting the preference, see below.)

Using the least cost method, A, with an EER of only 7, wins. Under the least cost with minimum energy efficiency method, C, the lowest priced model with an EER greater than 8, wins. Using percentage price differentials, D, which earns a $105 (35 percent of $300) preference for its 57.1 percent energy efficiency increase, wins. To summarize: the evaluation method selected determines which air conditioner manufacturer wins the bid. (This underlines the importance of setting out the evaluation method clearly in the invitation for bids.)

It is obviously necessary to have some way of evaluating the methods themselves. One way to test them is to compare the anticipated value of energy that would be saved in future years with the increased acquisition price [13]. Testing the validity of one's hypothetical example by further hypothecation is a bit incestuous, but it is intended for illustration only. Assuming that the air conditioner is to be used at full capacity for 1,000 hours per year, and that electricity costs 6 cents per kilowatt hour (kwh), one obtains the results set out in Table 1–2. (Note that the savings are not discounted to present value. If savings were discounted, payback periods would be somewhat longer. A discussion of discounting begins on page 22.)

As suggested by the illustrations, granting percentage price differentials to bids offering more energy-efficient commodities is a better energy conservation strategy than using energy-efficiency standards. Percentage price differentials facilitate trade-offs between purchase price and energy efficiency, while energy efficiency standards ignore the cost of individual units. Preferences are also more likely to keep pace with mechanical innovations, as they take a relative rather than an absolute approach to energy efficiency.

Establishing an Efficiency Differential. As already mentioned, the most complicated aspect of this strategy is establishing a ratio of price differentials to various increments of energy efficiency improvement. The ratio can be constant, or it can fluctuate in a way that rewards some increments more than others. The first requires a

Table 1-2. Hypothetical Time Required for Energy Cost Savings to Amortize Increased Acquisition Costs of More Energy-Efficient Air Conditioners *(derived from Table 1-1)*

EER	Watts*	Annual Energy Costs** ($)	Annual Energy Cost Savings as Compared to A's Energy Use ($)	Acquisition Cost Increase Over A ($)	Payback Period (years)
A 7	1,429	85.74	—	—	—
B 7	1,429	85.74	0	20	no payback***
C 9	1,111	66.66	19.08	80	4.2
D 11	909	54.54	31.20	100	3.2
E 12	833	49.98	35.76	140	3.9

Air Conditioners 10,000 BTUs.

  *Watts = (BTU) ÷ (EER).

  **Assuming 1,000 hours per year and 6 cents per kwh.

  ***Since there is no energy cost saving as compared to A's energy cost, there can be no payback.

fixed ratio, such as one to one (which means a 10 percent efficiency increase receives a 10 percent preference). The second approach requires that a scale be established, such as the one in Table 1-1.

Two things should be kept in mind when setting a scale: (1) the amount of energy, and hence money, a certain increase in efficiency is likely to save; and (2) the likely response of bidders to various levels of percentage differentials. Governments granting differentials will be concerned with the cost of the strategy and the likelihood that the cost will be recovered through energy savings over the operating life of the commodity. Further, it is important that differentials be granted in ratios that will motivate bidders to supply energy-efficient products and yet will cost no more than necessary to achieve this end.

If the energy efficiency of the lowest priced bid were known in advance, it would be easier to design the efficiency differential. If the low bid had a very low efficiency rating, then each percentage point of increased energy efficiency would be worth less than if the low bid had an average efficiency rating (1 percent of five is less than 1 percent of eight). Further, the cost and difficulty of improving energy efficiency is likely to increase geometrically and should be rewarded accordingly (e.g., a 30 percent increase is likely to be more than twice as costly to the developer as a 15 percent increase). Obviously, the low bid's efficiency rating cannot be known in advance if the contract is to be let by formal competitive bidding, so the differential must be an educated guess. For this reason, differentials are best set by individual purchasing agencies and by officials who have a feeling for the likely range of bid prices and efficiency ratings.

Conceptually, the granting of percentage price differentials falls between energy efficiency standards and life cycle costing as an energy conservation strategy. PPD suffers from being much more complicated to administer than efficiency standards and yet not nearly so accurate, in relating price increases to energy savings, as life cycle costing. Percentage price differentials do not constitute a highly recommended strategy; they are presented here merely as an alternative.

LIFE CYCLE COSTING

Problem Focus

Government procurement of buildings and commodities is often plagued by the "minimum first cost syndrome" [14] : competing bids are evaluated solely on the basis of stated acquisition price. Operating, maintenance, and other costs of ownership are thus ignored.

This unfortunate practice has a legal basis in state purchasing statutes that require contracts to be awarded to the "lowest responsible bidder" [15] . But state courts, with some exceptions, have allowed contract awards to higher bidders when they promise better quality goods or services [16]. This latitude allows purchasing agencies to buy energy-efficient items, even if they are not the cheapest.

Unfortunately, this opportunity is seldom pursued, as it is much easier to scan a list tabulating bids and simply pick out the lowest than to analyze each bid to determine energy consumption and other operating costs. Harried purchasing officials do not often have time to burn the midnight oil in search of statutory loopholes. Thus, amending the purchasing statute may be the only way to assure that energy cost becomes a factor in bid evaluation. Note, however, that the potential for putting life cycle costing into effect already exists under most existing purchasing statutes.

Strategy

Life cycle costing (LCC) is a general name for procurement techniques that consider operating, maintenance, and other costs of ownership, as well as acquisition price. Because energy expenditures constitute an increasingly large portion of the operating costs of many items, LCC represents significant energy conservation potential.

Modified Life Cycle Costing.   The somewhat complex discussion of total life cycle costing that makes up the bulk of this subchapter is intended to provide the reader with a certain depth of knowledge on the subject. It is *not* intended to convince small jurisdictions that LCC is "out of their league" as a purchasing strategy. By way of avoiding this latter contingency, the more complex discussion is prefaced by the presentation of a simpler form of LCC.

This simplified strategy, often referred to as modified life cycle costing, calls for considering only two costs of ownership—purchase price and energy consumption. The fact that this involves consideration of only one more cost than is evaluated under the "minimum first cost syndrome" possesses about the same significance as the fact that $212°F$ is only one degree more than $211°F$. Properly applied, evaluation of bids to supply energy-consuming goods on the basis of these two costs yields almost the entire energy conservation benefit potentially available through life cycle costing. "Proper" application implies that since modified LCC does not take such costs as maintenance and salvage into account, it should not be implied that these additional costs are expected to vary widely among competing products.

The following is a formula for modified LCC applied by the division of purchasing of the state of Florida:

Evaluation will be based on the lowest unit bid price plus operational cost over a 5 year period, in accordance with the following formula:

$$A = \frac{MCT}{E} + B$$

A = Award figure—lowest "A" wins award
B = Bid price
C = 4.0×10^{-5} (cost of electricity per watt hour)
E = Energy efficiency ratio
M = Minimum BTUs listed in specification for the air conditioner BTU range bid
T = 7500 hours (operational hours in a 5 year period)

Award will be on the lowest award figure on an item-by-item basis [17].

One improvement on the Florida formula, which would not complicate it, would be to consider operational cost over a longer period. This would be valid so long as the period did not exceed anticipated useful life of the machine, and lengthening the period would increase the formula's emphasis on energy efficiency. (A suggested legislative approach to implementing this strategy is found in the appendix to this chapter.)

Total Life Cycle Costing. The object of total life cycle costing (hereinafter LCC), as applied to public purchasing, is to make agencies aware of the comparative costs of acquiring, possessing, using, and disposing of a particular type of item. LCC requires that contracts be awarded to the bid that represents the lowest total cost of ownership. To apply LCC, a purchasing agency may elect to:

1. identify every cost that the government will incur as a result of owning the item (including acquisition costs and the cost of administering LCC analysis);
2. determine the useful life of the item;
3. estimate the salvage value of the item;
4. discount all costs and salvage value to present value;
5. subtract the salvage value, when identifiable, from the cost and divide the result by the number of years of useful life (this yields the *average yearly cost of ownership*);
6. compare the average yearly cost of ownership of the items under consideration;
7. buy the least expensive item or service.

A hypothetical application of LCC is presented in Table 1–3. It is simplistic, because it includes very few costs of ownership, holds energy cost constant over ten years, and does not discount future costs. (These omitted factors are examined later in this subchapter.)

It is obvious that LCC is much more than just an energy conservation strategy. First, LCC is cost-effective because it reduces the average yearly cost of ownership at the same time as it cuts energy consumption. It draws attention to wasteful operating costs that may have been obscured by emphasis on initial cost.

Second, LCC is a wise procurement technique, even in the absence of energy considerations, because it considers all costs rather than just energy costs. Relevant costs include maintenance and the expense of obtaining replacements during breakdowns. The useful life and the salvage value of the item also affect the cost of ownership. LCC, properly applied, reflects all such costs in a single figure.

Third, even a truncated version of LCC provides a better basis for comparing items than does reliance upon acquisition (first) cost. If all the costs of ownership cannot be quantified, using those that can is more valid than using only first cost. For example, even if it is only possible to identify maintenance costs, LCC analysis may demonstrate the financial wisdom of procuring an item that costs more initially if that item would have significantly lower maintenance expenses [18].

Table 1–3.   A Hypothetical Application of LCC

Assume: 1. Item A and Item B are offered in different bids. Both perform identically.
2. Electricity costs 5 cents per kwh.

Comparison:	Item A	Item B
Annual kilowatt hour consumption	1,000	3,000
Annual maintenance cost	$100	$100
Useful life	10 years	10 years
Price	$2,700	$2,200
Anticipated salvage value	$200	$200

Costs Over Ten Years: (undiscounted)

A	1	2	3	4	5	6	7	8	9	10	ten year total
Acquisition Cost	2,700										2,700
Energy Cost	50	50	50	50	50	50	50	50	50	50	500
Maintenance	100	100	100	100	100	100	100	100	100	100	1,000
Salvage										(200)	-200
Total Annual Cash Flow	2,850	150	150	150	150	150	150	150	150	(50)	4,000

B											
Acquisition Cost	2,200										2,200
Energy Cost	150	150	150	150	150	150	150	150	150	150	1,500
Maintenance	100	100	100	100	100	100	100	100	100	100	1,000
Salvage										(200)	-200
Total Annual Cash Flow	2,450	250	250	250	250	250	250	250	250	50	4,500

Results: Over their useful lives, item A would cost $4,000 to own; item B would cost $4,500. Thus, item A, even with an acquisition cost $500 higher than item B, has a lower cost of ownership over its useful life. Note that costs are not discounted to present value.

In the preceding section, three methods of evaluating bids for a hypothetical contract to supply air conditioners were compared: least cost, energy efficiency standards, and percentage price differentials. Percentage price differentials were found to be better than energy efficiency standards for balancing energy efficiency and purchase price and also for conserving energy. LCC, however, is even better. This assertion finds preliminary support in the application of LCC to the example in the preceding section. The illustration in Table 1–2 of using energy cost savings to pay back increased acquisition cost is analogous to the LCC concept. Analysis of the table suggests that the percentage price differential method may be superior, as it selected air conditioner D, the appliance that was the second most energy-efficient (EER of 11) and that paid back its increased acquisition cost ($100 increase over A) most rapidly (in 3.2 years). Percentage price differential, however, does not consider the useful life or the salvage value of competing items. In fact, it fails to consider any costs other than acquisition and energy costs. Consequently, to apply LCC to the hypothetical situation it is necessary to assume that all other costs, such as maintenance costs, are the same for all five models.

LCC will be applied in two examples: where useful life is assumed to be five years and where it is assumed to be nil, and future costs will not be discounted to present value. In Table 1–4, where useful life is only five years, the preference approach and LCC select the same air conditioner. But in Table 1–5, where useful life is ten years, LCC selects the more efficient air conditioner, E, even though it is significantly higher priced ($340 as compared to $300) than D. As the period of useful life of the appliance is not included, the PPD method fails in Table 1–5 to make the most cost-effective choice.

In terms of energy use, the results are: in Table 1–4, both LCC and the PPD approach trade 385 kwh of electricity for the $40 saving of purchasing D rather than E. In Table 1–5, LCC finds cost justification for saving 770 kwh that the preference approach does not save.

Before the reader eagerly accepts LCC as the universal antidote, however, it must be pointed out that LCC is a complex strategy, involving more difficult and expensive administration. The major complexities, glossed over thus far, evolve from the necessary steps of (1) determining what costs to include, (2) projecting prices into the future, and (3) discounting costs to present value. Fairness among bidders obviously requires that these determinations precede, and be set out in, the bid invitation. Decisions in all three categories may ultimately determine which bidder wins the contract, and bid-

Table 1-4. LCC v. Percentage Price Differentials—5 Years *(Derived in part from Tables 1-1 and 1-2)*

	Cost of Acquisition	Yearly Energy Cost	Useful Life (yrs.)	Total Cost of Ownership	Average Yearly Cost of Ownership
A	$200	$85.74	5	$628.70	$125.74
B	220	85.74	5	648.70	129.74
C	280	66.67	5	613.35	122.67
D*	300	54.54	5	572.70	114.54*
E	340	49.98	5	589.90	117.98

*Lowest average yearly cost.

Table 1-5. LCC v. Percentage Price Differentials—10 Years *(Derived in part from Tables 1-1 and 1-2)*

	Cost of Acquisition	Yearly Energy Cost	Useful Life (yrs.)	Total Cost of Ownership	Average Yearly Cost of Ownership
A	$200	$85.74	10	$1,057.40	$105.74
B	220	85.74	10	1,077.40	107.74
C	280	66.67	10	946.70	94.67
D	300	54.54	10	845.40	84.54
E*	340	49.98	10	839.80	83.98*

*Lowest average yearly cost.

ders must all know what they are bidding for. Further, while every feasible effort should be made to include all relevant costs, such costs must not be speculative. Bid evaluation and contract award are administrative processes that are subject to judicial review, and if a purchasing agency fails to provide sound reasons for assigning costs to various bids, courts may overturn contract awards.

Determining What Costs to Include
If LCC is to provide an accurate reflection of the cost to the government of owning a particular item, a great many costs must be considered that are not ordinarily thought of. These include costs of performing the LCC analysis, integrating the item into government use, providing any other items that are necessary for the first item to function (e.g., computer cards for a computer), and transporting the item (if the government pays the bill).

The following list of costs was adopted from a manual on LCC by the U.S. General Services Administration [19]:

Source selection costs
  Special surveys
  Test procedures
  Additional employees

Acquisition costs
  Development
  Hardware
  Special ancillary equipment
  Installation

Support
  Operation
    Employees
    Energy

  Maintenance
    Parts
    Employees
    Transportation

  Inventory Management

Training
  Maintenance
  Hardware
  Employees
  Training aids
  Operation

Inspection and acceptance
  Hardware
  Employees

Transportation

Documentation
  Drawings
  Manuals
  Parts lists
  Specifications

Disposal
  Net scrapping cost
  (Resale value)

The length of this list suggests that LCC requires substantially more analysis than energy efficiency standards or percentage price differentials. Obviously, some of these costs would only be encountered in the acquisition of fairly complex items. But in any application of LCC, it should be remembered that estimates must have a sound basis if the contract award is to survive judicial scrutiny. Second, an inability to determine some costs need not prevent the application of LCC. As asserted on page 16, even the simplistic application of LCC is a better strategy than relying solely on minimum first cost.

Once a purchasing agency determines which costs to include in LCC analysis for a particular procurement, the agency must attach a figure to each cost. There are at least two ways this could be done. One way is for the purchasing agency to test the products submitted to determine their useful life and their maintenance and operating costs. The agency would, of course, keep a record of its tests and results for potential use in court. The administrative costs of this method would be high. Also, it is not possible to determine useful life of certain products by short term testing.

At the other extreme, the bidders or other potential suppliers could perform the costing and warrant its accuracy in the contract. This would appear to shift the burden to the bidders, but the government will actually have to pay for the warranties in either case— whether directly or through higher prices. While this Hobson's choice is unappetizing, there is some middle ground. The Energy Policy and

Conservation Act of 1975 mandates federal testing and labeling of the energy efficiency of major appliances and automobiles [20]. Once instituted, this rating process could readily be used in LCC analysis. In the interim, while waiting for the administrative process to grind to conclusion, state and local governments can rely to a limited extent upon the voluntary labeling program in which a few manufacturers participate [21]. Governments should make an effort to communicate results of their own tests to other jurisdictions.

The U.S. General Services Administration (GSA) recently applied LCC to the procurement of five types of room air conditioners [22]. The GSA's costing procedures were simple. Costs fell into three categories: acquisition, initial logistics, and recurring. The first category was simply the price for an air conditioner. Since the procurement was to be of commercially available items, initial logistics costs were negligible. The third category, recurring costs, consisted essentially of maintenance and operating costs. Warranties were used to cover maintenance costs, and GSA accepted the Association of Home Appliance Manufacturers' certification of air conditioner EER to determine the energy cost. Expected useful life was estimated at seven years. Obviously, costing need not be an onerous requirement in every instance.

Projecting Prices Into the Future

If the multitude of studies of energy supply and demand that have been performed since the winter of 1973 have shown anything, it is that it is not possible to accurately predict the price of electricity ten or fifteen years hence [23]. There is such a diversity of "expert" opinion that private consumers would do as well by making up their own projections.

Unhappily, however, purchasing agencies that make up statistics are likely to find themselves defending their creations in court. Few people doubt that, at least during this century, energy prices will continue to increase. Since there is little consensus as to how great this increase will be, or as to when it will occur, purchasing agencies may be wise to stick close to home, i.e., to use present energy prices in their areas for LCC analysis. Any projections of increased costs should be as well documented as possible in studies by venerable institutions.

Despite these legal perils, it is highly desirable, from the standpoint of energy conservation, to project operating cost increases—particularly where the item to be procured has a long useful life, such as a building or major appliance. Projected increases in operating costs can be used to justify additional expenditure for energy efficiency and durability [24]. Purchasing agencies are thus advised to see how close to the cliff they can walk without falling off.

Discounting Costs

Since LCC considers costs that are to be incurred in the future (such as operating and maintenance costs), it is necessary to discount these costs to their present value. This is because money in hand is worth more than money to be received in the future. Manifestations of this belief include the payment of interest on savings accounts and loans. Thus, a $10 increase in present costs (such as in the acquisition price of an air conditioner) must be justified by more than a $10 savings (such as in energy consumption) to be realized in the future. The longer the time between present expenditure and realization of future saving, the more the future saving must exceed the present expenditure.

Setting a discount rate is a policy decision. The percentage discount constitutes an expression of how much money an agency must receive in the future to be willing to incur a present cost. This decision is very important to the conservation of energy via increased energy efficiency. The lower the discount rate, the easier it is to justify increased expenditure for energy-efficient buildings and energy-consuming appliances. Governments seriously concerned with energy conservation should be willing to accept a relatively low dollar return on their "investments" in energy efficiency to achieve a higher rate of conservation. Since energy conservation represents a societal benefit independent of dollar cost considerations, purchasing agencies can set the discount rate at a minimum, break-even level and still have a high benefit to cost ratio.

The U.S. Office of Management and Budget (OMB) applies a 10 percent discount rate to all executive branch agencies of the federal government except the Postal Service [25]. OMB feels that this rate represents an "estimate of the average rate of return on private investment, before taxes and after inflation" [26]. The U.S. Environmental Protection Agency recently adopted a discount rate of 6.125 percent for use in cost-effectiveness analysis under the Federal Water Pollution Control Act Amendments of 1972 [27].

These examples are offered to provide perspective for readers unfamiliar with discounting, and are not intended as recommendations to state or local purchasing agencies.

Where operating costs are to be incurred at a constant rate over the useful life of an item, it is appropriate to use the discount factor sum. This is a single number, which is multiplied by the annual operating cost to yield the present value sum of the annual operating cost of the item over its expected life. If one multiplies the annual operating cost by the number of years of operation, the result is the undiscounted sum operating cost. But if the discount factor sum replaces the number of years, the result is the sum cost discounted to present value. The discount factor sum will always be a number that is less than the number of years. The formula for discount factor sum is:

$$\frac{1 - [1 - r(.01)] - n}{r(.01)}$$

r = discount rate
n = number of years

Where the anticipated period of use of an item is less than its usable life, or where an item can be recycled, LCC analysis should also include an item's salvage value. Since salvage value is realized only at the end of the period of use, a different formula must be applied to determine the present worth of the salvage value:

$$PV = \frac{SV}{(1 + r)^n}$$

PV = present value
SV = salvage value
r = discount rate
n = number of years

(The uninitiated may be relieved to know that tables are available that simplify these calculations [28].)

Present Applications of LCC
to Evaluating Competitive Bids

A number of states have bills before their legislatures to require the application of LCC to certain government acquisitions; a few already have laws requiring it [29]. Sample laws are reproduced in the appendix. In addition, there are several such bills before the second session of the Ninety-fourth Congress [30]. LCC is gaining in popularity, as evidenced by its recent endorsement by the energy task force of the National Council of State Legislatures [31], but its application is not so widespread as might be expected in light of its great potential for saving energy and money.

The Federal Supply Service (FSS) of the GSA recently awarded two supply contracts, one for room air conditioners and one for water heaters, on the basis of LCC. The FSS documentation of the water heater procurement is an excellent case study, and a brief summary is valuable as a final illustration of the application of LCC to competitive bidding [32].

The water heater procurement was initiated by a formally advertised invitation for bids. Two bids were submitted and the contract
was awarded to the bid reflecting the lowest total cost of ownership.
Previous FSS procurements of water heaters had been awarded to
the bid with the lowest price, without regard to the total cost. The
only procedure that might be foreign to some state purchasing agencies was the requirement that technical proposals be submitted prior
to bids to allow the agency to check the adequacy and suitability
of the engineering and its conformance to the specifications. All
technical proposals submitted were deemed acceptable.

As with the air conditioner procurement mentioned earlier, expenses were divided into acquisition, initial logistics, and recurring
costs. Acquisition cost was simply the unit price per heater; initial
logistics costs were found negligible. As to recurring costs, maintenance expenditures were eliminated from consideration as there were
warranties leaving only operating costs (energy). The energy efficiency of water heaters is slightly more difficult to compute than
that of air conditioners, because the amount of heat transferred to
the water (thermal efficiency) must be reduced by the amount of
heat that is lost during water storage (standby loss). Once again, however, the FSS was able to use performance standards developed by
the industry.

The following LCC formula was used:

$$L = X + Y$$

L = life cycle cost (present value dollars)
X = acquisition cost
Y = present value sum of the annual operating
cost of the product over its expected life [33].

Salvage value was apparently considered to be nil, since it is not mentioned, and product life was set at ten years for all bids [34].

Each of the two competing suppliers submitted bids, on four types
of heaters. From a case study perspective, the results were classic.
Different bid prices were permitted for each of ten geographic areas,
thus taking variations in transportation and energy costs into account.
As a result, there were forty discrete bid prices and LCC compari-
sons. In *every case* bidder A's prices were lower than bidder B's, and
in *every case* B's life cycle cost (discounted to present value) was
lower than A's. Based on the anticipated purchase of 7,650 heaters,
the present value of anticipated cost savings totaled $326,457.00
[35]. While anticipated energy savings were not included they work
out to the rough equivalent of 79.4×10^6 kwh or 58.7×10^3 barrels
of natural gasoline over ten years.

FSS evaluated bids in this case study on the basis of total life cycle cost, rather than the average yearly cost of ownership as recommended in this book. For maximum savings of both cost *and energy*, LCC analysis should be based on the average yearly cost of ownership rather than total life cycle cost [36]. The average yearly cost of ownership is obtained by dividing the total life cycle cost by the anticipated useful life. It is usually illogical to assume that the useful lives of competing products are equal, but the difficulty of predicting useful life is a substantial impediment. One possible way out is to set useful life as a requirement.

The FSS failure to consider salvage value is another minor shortcoming. Unless the product "self-destructs" at the end of its anticipated useful life, it will have some salvage value. Even if the item no longer functions, its components can be recycled to conserve energy and other natural resources [37]. In many cases, however, salvage value will be about the same for all competing items, so that it can be cancelled out of the LCC equation.

Evaluation of Alternative Designs
for Government Buildings Prior
to Soliciting Bids

The previous sections discuss ways of applying LCC to competitive bidding. We now turn to how several states are applying LCC to the evaluation of building designs. The average modern government building has many interrelated energy-consuming components [38]. In addition, building components that do not consume energy, such as windows and walls, help determine a building's energy efficiency. LCC analysis for a building is thus far more complex than for an air conditioner. And as government buildings are usually designed before bids for construction contracts are solicited, LCC analysis must therefore occur during the design stage.

Florida and Washington were the first states to respond to this need. Both states passed laws requiring the agency responsible for constructing a building to include a life cycle cost analysis in the design phase [39]. (These laws are reproduced in the appendix.) Under these statutes, LCC analysis performs a sort of advisory function. The Washington law requires a "comparison of three or more energy system alternatives" [40]. It makes no specific reference to LCC analysis as a *mechanism* for selecting the energy system design: "A public agency may accept the facility design if the agency is satisfied that the life-cycle cost analysis provides for an efficient energy system or systems based on the economic life of the major facility" [41].

A memorandum of the department of general administration (DGA) of the state of Washington sheds some additional light on the function of LCC in design selection. The memorandum requires that LCC analysis include:

a. Summary of Owner and Operating cost on a present worth basis.
b. Summary of Owner and Operating cost on an equivalent annual cost basis.
c. A payback or break-even cost benefit comparison of the three (3) mechanical system alternatives.
d. A summary outline of the rationale for selection of final alternatives of architectural variables and mechanical system employed in determining the proposed recommendation [42].

LCC analysis in Washington is thus the second line of defense against energy waste. First, the building is designed according to standards promulgated by the DGA [43]. Then LCC suggests which design provides the optimal ratio of energy efficiency to acquisition or construction cost. This approach is sound, provided the energy conservation standards are well conceived.

Implementation of the Florida law seems more complicated at first, but is essentially the same [44]. It requires at least two alternative energy consumption system designs [45]. Alternative designs must stay within the energy budgets established by the Florida Department of General Services (DGS) for the type of building [46]. The alternative designs are run through a computer program, the Florida Lifecycle Energy Evaluation Technique (FLEET), which yields both their energy performance indices and life cycle analyses [47]. Each alternative must reduce building energy consumption below the level set by the DGS, and LCC analysis selects the alternative that achieves the greatest reduction. As numeric energy efficiency standards tend to stagnate (see p. 7), the Florida method represents an innovative strategy for overcoming the problem.

Another exposition of the application of LCC to government building construction is in *Energy Conservation Design Guidelines for New Office Buildings*, a manual prepared for the GSA [48]. This brief discussion presents interesting approaches to several problems of LCC application. It is reproduced in the appendix (p. 47).

Legal Implementation and Stumbling Blocks to Applying LCC to Procurement of Both Commodities and Buildings

While the limited application of LCC by the Federal Supply Service preceded a congressional mandate (which appears to be forthcoming), universal application of LCC by state and local purchasing agencies is unlikely because of the relative ease of selecting the product or building with the lowest initial cost [49]. Hence, where the legislature senses that administrative action is not likely, or where the existing purchasing statute is so rigid as to actually prohibit LCC, it may wish to enact legislation requiring its implementation. Please note, though, that in many instances a change in the purchasing law is not required, and legislative inaction should not be viewed, from the agency level, as an impediment. This fact is borne out by the Florida experience of modified LCC (See p. 15).

Likely legal or institutional stumbling blocks include difficulties related to: (1) higher acquisition prices, (2) projecting future prices, and (3) the restriction of competition. The problem of projecting prices has already been discussed on pages 21−22.

A higher acquisition price may present problems, even if an item has a lower operating cost and will last longer than alternatives, because a government may just not want to spend money at the particular time. The public officials who implement a full-scale LCC program may have to raise taxes, at least in the short run. For obvious political reasons, they would prefer to avoid doing so. When legislatures appropriate meager lump sums for a project, it reinforces the "minimum first cost syndrome." Procurement officials must, therefore, educate legislators as well as the public, emphasizing how more generous appropriations may substantially reduce operation and maintenance costs. In the state of Washington, the trend is for the legislature to initially appropriate only the design costs of new state buildings; construction costs are separately appropriated after the LCC analysis is complete [50].

To the extent that a purchasing agency is allowed to incur indebtedness, it can include in LCC analysis the interest payments necessitated by borrowing money to pay higher acquisition prices [51]. Thus, a paucity of available funds would no longer be an excuse for not applying LCC, and the savings in operating costs would pay off the loan, with interest.

LCC analysis may be challenged on the grounds that it is detrimental to competition—i.e., that the categories of costs selected for inclusion in the analysis may help or hinder the various bidders. While the possibility for such criticism should motivate procurement officials to ensure that chosen costs accurately reflect reality, it is not really a valid criticism of LCC. As has been amply demonstrated, LCC facilitates competition by allowing product comparison on the basis of many costs rather than just purchase price.

Advantages and Disadvantages of LCC as an Energy Conservation Strategy in State and Local Procurement

The following checklists summarize the benefits and shortcomings of LCC.

Advantages

1. LCC is cost-effective.
2. LCC is a wise procurement technique even in the absence of energy conservation considerations.
3. LCC can be applied in an abbreviated form and still provide a better comparison of alternative acquisitions than is provided by low first cost.
4. LCC provides a more effective trade-off between acquisition price and energy (operating) costs than either efficiency standards or percentage price differentials.

Disadvantages

1. LCC requires more complex testing and other analysis and may be more expensive to apply than either energy efficiency standards or percentage price differentials.
2. Legislation may be required to implement LCC.
3. LCC may require payment of higher initial costs [52].

DESIGN VERSUS PERFORMANCE SPECIFICATIONS

Problem Focus

It is more likely that suppliers of energy-consuming products will be able to improve the efficiency of these products if there are few restrictions on product design. Where product design is fully controlled by government specifications, suppliers may be unable to innovate.

Strategy

Design specifications describe the way a product must be constructed. They are roughly equivalent to blueprints. Performance specifications describe the way a product must perform; the product may be constructed in any way imaginable, and of any materials the contractor deems suitable [53].

Most specifications are combinations of performance and design. A greater emphasis on performance, rather than design, offers more opportunity for improving energy efficiency.

VALUE INCENTIVE CLAUSE

Problem Focus

Manufacturers who supply state and local governments with energy-consuming commodities are usually in a better position than the governments themselves to improve the energy efficiency of the commodities. Manufacturers are not likely to make this effort gratuitously, particularly if they have already been awarded a fixed price supply contract. It is necessary for governments to find a way to encourage manufacturers to improve the efficiency of their product during the life of the contract.

Strategy

The value incentive clause (VIC) is a procurement strategy implemented by the U.S. General Services Administration (GSA) in April of 1975 [54]. GSA requires that the clause be included in certain fixed price supply contracts that exceed, or are expected to exceed, $100,000. The VIC encourages contractors, during the life of the contract, to submit value change proposals (VCPs):

> VCP's contemplated are those that would result in net savings to the Government by providing either (1) a decrease in the cost of performance of this contract, or; (2) a reduction in the cost of ownership (. . . collateral [including operating] costs . . .) of the work provided by this contract, regardless of acquisition costs [55].

Obviously, these contracts must be governed by *design*, rather than *performance* specifications. Under performance specifications, the contractor determines the initial design and could thus deliberately design opportunities for subsequent value changes. If a VCP is accepted, the contractor is entitled to share in savings that are attributable to the value change [56]. Under the present GSA version, the contractor's share can be up to 50 percent of the savings [57].

The VIC has obvious energy conservation potential [58]. As has been repeatedly stressed in this chapter, energy is a major operating cost of energy-consuming commodities. The VIC therefore encourages manufacturers to improve the efficiency of their commodities.

Implementation and Impediments

Since it may take considerable time for a contractor to respond to the VIC by submitting a VCP, and since this submission must take place during the life of the contract, the VIC strategy is best suited to long term contracts. This is a major impediment to the effective implementation of this strategy, as many states and localities have laws restricting the duration of government contracts [59].

There do not appear to be any other significant obstacles to implementing this strategy. If a legislature wishes to take the initiative, it may do so, but this is not essential. A purchasing agency could probably implement a value incentive clause by issuing a regulation.

Unless administrative costs prove burdensome, this strategy should produce an excellent benefit to cost ratio; payments to the contractor come entirely from savings attributable to the contractor's effort.

ENCOURAGING SUBMISSION OF UNSOLICITED PROPOSALS

Problem Focus

Improving the energy efficiency of state and local government operations is part of a larger national problem—applying recent technological advances to the improvement of government operations [60]. The first step in the solution of any problem is perceiving that it exists. Lacking sophisticated, up-to-date knowledge of available technology, state and local officials may be unaware of the need for change, let alone how to go about it. It is unrealistic to expect officials to perceive the need for a particular technology and then seek out a supplier. In many instances, developers and suppliers have better information, and they often approach governments with suggestions. This may be so even where state or local officials have resources for technology evaluation. It is wise to encourage developers and suppliers of energy-efficient technology to participate in the perception and solution of these government problems.

Strategy

Publicizing a government's eagerness to receive unsolicited proposals for saving energy in government operations and promptly evaluating such proposals would give suppliers a significant incentive. Contracts would be awarded to the submitters of meritorious proposals.

A caveat must precede further discussion: Since this process would result in negotiated, noncompetitive procurement, it may run afoul of requirements that contracts be let only after competitive bidding. Nearly all states and many local governments have such a requirement [61]. Two things can be done to avoid this shoal. First, the purchasing agency could attempt to justify the deviation from competitive bidding by demonstrating that the supplier is the only one capable of furnishing the product or service. This type of justification, called "single source," would permit waiver of the competitive bid requirements of many purchasing statutes [62]. Second, the legislature could amend the purchasing statute to permit noncompetitive procurement in this instance. In addition, it should be noted that in many jurisdictions this strategy would be characterized as a research and development (R&D) method. As such, it would be foreign to many purchasing agencies.

If these legal hurdles can be overcome, implementation of this strategy becomes primarily a public relations effort. Governments must convince suppliers that proposals that achieve the stated objective at a reasonable cost will be accepted. The more strongly this belief is reinforced, the more likely it is that suppliers will examine government needs and submit proposals.

DESIGNATION OF SPECIAL PURPOSE (SPECIFICATIONS)

Problem Focus

In certain instances, purchasing agencies may wish to use design specifications to procure energy-consuming commodities, and, at the same time, allow for design alterations to improve energy efficiency. (For a discussion of design and performance specifications, see page 28.)

Strategy

Designating a special purpose is a procurement technique used where specialized items, such as sewing machines for the blind, are required [63]. In such instances, bids can be solicited on the basis of the ordinary design specifications for sewing machines, along with requests for proposals of how to adapt them to the specialized need.

The same could be done for energy-consuming commodities— bidders could be asked to submit proposals of how to change the specifications to improve the item's energy efficiency. For example, in procurement of refrigerators, suppliers might propose to better insulate the walls. This, in turn, might allow the use of a smaller motor. The acquisition price may fall, while the energy efficiency rises.

Implementation

It would seem that this fairly simple strategy could be successfully implemented by the purchasing agency, without legislative mandate.

NEGOTIATION AND REVISION OF COST

Problem Focus and Strategy

A recent report by the U.S. Environmental Protection Agency on solid waste makes the following observations about state competitive bidding laws:

"Lowest responsible bidder" and other procurement laws were instituted for public-sector purchase of risk-free, off-the-shelf technology. They do not work particularly well when applied to the procurement of a somewhat risky resource recovery system because such a purchase, by its nature, almost [always] requires a negotiating period before final signing of the contract. During the negotiation period, risk apportionment (between the public and private sector) and specifications for the products and waste processing can be determined. Unfortunately, reallocation of risk alters costs. Competitive bidding requirements usually make negotiations and revisions of cost illegal [64].

These observations are very relevant to the procurement of a complex, energy-consuming system like the heating, cooling, and ventilation of a government office building. Given the long useful life of a building, it is important that energy-efficient technology be incorporated during construction. A small increment of energy efficiency can make a big difference in operating costs over forty years.

To obtain the most advanced energy-efficient technology available at the time a building is constructed, it may be necessary for a government to engage in long negotiations with a single contractor [65]. During this period, the contractor may design various systems, have them evaluated by the government, partially redesign them, and install them. If the systems are to be on the cutting edge of the relevant technology, their cost will probably be uncertain during most of this period and will require revision [66].

Implementation

The legislature should amend purchasing statutes to allow for negotiation and revision of costs in the procurement of high technology, energy-consuming items. (Since the problem here presented applies to procurement of many types of complex systems, etc., the legislature might want to study what other types of procurements are involved and address the problem in a comprehensive fashion.)

CONCLUSION

It is hoped that state and local purchasing agencies will find ways to integrate appropriate strategies into their procedures. Particular attention should be paid to the concepts of life cycle costing and the use of performance specifications in the procurement of energy-consuming items.

A number of strategies presented in the chapter, such as life cycle costing, value incentive clauses, and performance specifications, would stimulate the development of energy-efficient technology by the private sector. The experimental technology incentives program of the U.S. National Bureau of Standards utilizes state and local procurement to stimulate technology development and its application to governmental needs. This program will provide valuable information and guidance to interested governments [67].

Increasing the energy efficiency of government operation should be a continuing goal, and an integral part of government procurement policy. Governments that recognize this today will be better prepared to cope with the energy shortages that may occur in the coming decades.

NOTES TO CHAPTER 1

1. *See* Environmental Law Institute, Energy Conservation Project, *ECP Report*, no. 2 (Washington, D.C., October 1975) and no. 4 (Washington, D.C., January 1976). These copies of *ECP Report* include a survey of pending and enacted state energy legislation.

2. Pub. L. No. 94–163 (1975), 42 U.S.C. §§ 6201 et seq.

3. 42 U.S.C. § 6322(c)(3).

4. For further information contact: F. Trowbridge vom Baur, Chairman, Coordinating Committee on a Model Procurement Code, 1700 K Street, N.W., Washington, D.C. 20006; telephone (202) 331–0133. See also, vom Baur, The Project for a Model Procurement Code, 8 *Public Contract Law Journal* 4 (1976).

5. *See* Clean Air Amendments of 1970, § 306, 42 U.S.C. § 1857h-4 (1970), ELR 41225; Federal Water Pollution Control Act Amendments of 1972, § 508, 33 U.S.C. § 1368 (Supp. 1973), ELR 41125.

6. For air conditioners, energy efficiency is expressed in terms of their energy efficiency ratio (EER). This is a measure of the number of BTUs of cooling power produced by each watt of electricity consumed. The higher the EER, the greater the efficiency of the air conditioner. Note that according to the Association of Home Appliance Manufacturers (AHAM), reliable testing methods for appliance energy efficiency are presently available for air conditioners, refrigerators, combination regrigerator-freezers, and freezers. Methods for testing ranges and dishwashers are currently in the final stages of development. (Letter from Mr. Frank Miles of AHAM, October 19, 1976.) Note further that Sections 322 and 323 of the Energy Policy and Conservation Act of 1975, 42 U.S.C. §§ 6292 and 6293 (1975), require that the U.S. National Bureau of Standards develop testing methods for at least thirteen energy-consuming products:

refrigerators and refrigerator-freezers; freezers; dishwashers; clothes dryers; water heaters; room air conditioners; home heating equipment, not including furnaces; television sets; kitchen ranges and ovens; clothes washers; humidifiers and dehumidifiers; central air conditioners; furnaces; any other type of consumer product which the Administrator, of the Federal Energy Administration, classifies as a covered product. . . .

42 U.S.C.§6292(a). It is thus likely that a broad range of testing methods is forthcoming.

7. *See* Appendix, p. 42. None of the bills extracted here have been enacted into law.

8. "Author's Suggested Legislative Approach Number One," relating to the energy efficiency of government vehicles, is found in the appendix to this chapter.

9. "Author's Suggested Legislative Approach Number Two," relating to the energy efficiency of government vehicles, is found in the appendix to this chapter.

10. Motor Vehicle Information and Cost Savings Act, §510(a), 15 U.S.C. §2010(a), created by the Energy Policy and Conservation Act of 1975, 42 U.S.C. §6201 et seq. Emphasis added.

11. For example, two federal acts require that higher prices be paid for quiet machines and low pollutant-emitting motor vehicles. *See* Section 15(c) (1) of the Noise Control Act of 1972, Pub. L. 92−574, 42 U.S.C. §4914(c) (1), ELR 41506; and Section 212(d) and (f) of the Clean Air Act, *as amended*, 42 U.S.C. §1857f−6e(e) and (f), ELR 41215−16. The price differential may be set as high as 200 percent in order to spur private development of "clean" motor vehicles. *See also* Environmental Law Institute, *Federal Environment Law*, ed. Erica L. Dolgin and Thomas P. Guilbert (St. Paul, Minnesota: West, 1974), pp. 473−75.

12. The Council of State Governments, *State and Local Government Purchasing* (Lexington, Kentucky: The Council of State Governments, 1975), p. 9.4. Note also that the council soundly condemns this practice as tending to encourage in-state bidders to submit artificially high bids and otherwise to inhibit competition for government contracts. This criticism is not applicable to the granting of preferences to energy-efficient commodities. While it is true that price competition may be inhibited, this will be so only because undesirable, energy-wasteful commodities will receive a handicap. In a broader sense, competition will be enhanced, because it will be on the basis of operating cost (energy consumption) as well as acquisition price.

13. This test is analogous to life cycle costing (LCC), the strategy presented in the next subchapter. As will be discussed there, LCC provides a more accurate trade-off between price and energy efficiency than is provided by either energy efficiency standards or percentage price differentials. Total LCC is somewhat more complex to administer, but modified LCC is probably easier to use than PPDs.

14. *See e.g.*, *An Analysis of the ERDA Plan and Program*, U.S. Congress, Office of Technology Assessment, October 1975, p. 191. This report identifies the minimum first cost syndrome as a primary nontechnical barrier to energy conservation in buildings.

15. *See e.g.*, Cal. Gov't Code §14330 (West 1963); Cal. Gov't Code §14807 (West Supp. 1976); Cal. Gov't Code §25454 (West 1968); Minn. Stat. Ann. §1608 (Supp. 1976); N.Y. Gen. Munic. Law §103 (McKinney Supp. 1975); Wash. Rev. Code Ann. §35.23.353 (1975); and Wyo. Stat. Ann. §15.1–13 (Supp. 1975).

16. *See* Annot. 27 A.L.R.2d 917 (1953).

17. State of Florida, Department of General Services, Division of Purchasing, Specifications for Air Conditioners; Electric, Motor Driven, Self Contained, Window Mounted, No. 030–04, effective June 24, 1976. Note that this modified LCC formula has been adopted and implemented under the existing purchasing statute, which makes no specific reference to LCC.

18. On the other hand, LCC analysis might demonstrate that the increased cost of maintenance would be compensated for by savings in acquisition price.

19. Federal Supply Service, U.S. General Services Administration, Life Cycle Cost Workbook (Washington, D.C., n.d.), p. I–7.

20. Title V, Motor Vehicle Information and Cost Savings Act, 15 U.S.C. §§ 2001–2012, created by the Energy Policy and Conservation Act of 1975, 42 U.S.C. §6201 et seq.; and Sec. 324, Energy Policy and Conservation Act of 1975, 42 U.S.C. §6294.

21. *See e.g.*, "Voluntary Energy Conservation Specification No. 1–74, for Room Air Conditioners," U.S. Department of Commerce, 39 Fed. Reg. 15197 (1974).

22. *See* Federal Supply Service, U.S. General Services Administration, *Life Cycle Costing in the Procurement of Room Air Conditioners: LCC Procurement Case 1* (Washington, D.C., July 1975).

23. *See e.g.*, U.S. Federal Power Commission, *The 1970 National Power Survey, Part 1* (Washington, D.C.: GPO, 1971), pp. I–19–1–I–19–11.

24. Note that items that are more durable are big energy savers as well as items that are energy-efficient. The amount of energy required in the production of virtually any item is enormous, given the present high level of energy intensiveness in U.S. manufacturing processes. Thus, durable items save energy by putting off the need for their replacement.

25. *See* Office of Management and Budget Circular No. A–94, revised March 17, 1972.

26. *Ibid.*, p. 4.

27. 41 Fed. Reg. 15446 (April 13, 1976).

28. There are literally dozens of publications that include these tables. Copies of Union Carbide's booklet, *Compound Interest Tables*, may be obtained by writing to: Public Relations Department, Union Carbide Corporation, 270 Park Avenue, New York, N.Y. 10017.

29. *See e.g.*, Cal. Gov't Code §14951 (West Supp. 1975); Fla. Stat. Ann. §255.251 et seq. (1975); N.C. Gen. Stat. §§143.64.10–143.64.14 (1975); and Wash. Rev. Code Ann. 39.35.010 et seq. (1975).

30. S. 3217 by Mr. Gary Hart (introduced March 19, 1976), and S. 3240 by Mr. Bentsen (introduced April 1, 1976).

31. *See* National Council of State Legislatures, Energy Task Force, *Energy Report to the States* (Washington, D.C., March 19, 1976), pp. 4–5.

32. *See LCC Procurement Case 1, supra* note 18, p. 22; and Federal Supply Service, U.S. General Services Administration, *Life Cycle Costing in the Procurement of Water Heaters: LCC Procurement Case 2* (Washington, D.C., July 1975).

33. *LCC Procurement Case 2, Ibid.*, p. 3.

34. There is no indication in the documentation as to how this figure was arrived at or why it was applied equally to all bids and all models within each bid.

35. *LCC Procurement Case 2, supra* note 28, p. 17.

36. For discussion of the importance of product durability to energy conservation, *see* note 20, *supra.*

37. In many instances, recycling can produce enormous energy savings. For example, recycling aluminum saves 96 percent of the energy required to produce this metal from virgin ore. *See* William E. Franklin, David Bendersky, William R. Park, and Robert G. Hunt, "Potential Energy Conservation from Recycling Metals in Urban Solid Waste," in *The Energy Conservation Papers*, ed. Robert H. Williams (Cambridge, Massachusetts: Ballinger, 1975) p. 172.

38. For example, air conditioning systems may have to work harder to compensate for the heat produced by some lighting systems.

39. Fla. Stat. Ann. § 25551 et seq. (1975) and Wash. Rev. Code Ann. § 39. 35.010 et seq. (Supp. 1975). The Florida law applies to all state buildings; the Washington law applies only to buildings having 25,000 square feet or more of usable floor space.

40. Wash. Rev. Code Ann. § 39.35.030 (9) (a) (Supp. 1975).

41. *Id.* § 39.35.040.

42. Energy Management Office, Department of Administration, State of Washington, Memorandum, "Guidelines for Preparation of Life Cycle Cost Analysis for Publicly Owned or Leased Facilities" (Olympia, Washington, September 1, 1975). (Emphasis added.)

43. *See* Department of General Administration, State of Washington, Memorandum, "Energy Conservation Standards for State Facilities" (Olympia, Washington, July 18, 1975).

44. This explanation is based upon: *The Florida Life Cycle Analysis Manual* and the *Florida Energy Conservation Manual*, both prepared for the Florida Department of General Services (DGS) by PGA Engineers, Inc. of Tampa, Florida (Consultant: Syska and Hennessy, Inc. of New York City), March 1975. Additional information source is telephone interview of Thomas A. Sechler, Administrator of the Bureau of Construction, DGS, by Richard Reeve, a research assistant at ELI.

45. Florida Stat. Ann. § 255.255(5) (e) (1) (1975).

46. *See Florida Energy Conservation Manual, supra* note 40, Chapter 2.

47. *Florida Life Cycle Analysis Manual, supra* note 40, p. 5.

48. Public Buildings Service, U.S. General Services Administration, *Energy Conservation Design Guidelines for New Office Buildings*, 2nd ed. (Washington, D.C., July 1975).

49. The Federal Supply Service effort was not without some external encouragement, in the form of a U.S. General Accounting Office Report, "Ways to Make Greater Use of the Life Cycle Costing Acquisition Technique in DOD [Department of Defense]," issued on May 21, 1973. This report, in addition to

making the recommendation indicated by its title, stated that civilian agencies could benefit by using LCC. The apparently forthcoming congressional mandate is that of two Senate bills, S.3217 and S.3240. *See* note 26, *supra.*

50. Interview with Raymond Anderson of the Washington State Department of General Administration.

51. *See* Advisory Commission on Intergovernmental Relations (ACIR), *State Constitutional and Statutory Restrictions on Local Government Debt* (Washington, D.C., 1961).

52. Recent studies have shown that, at least as to buildings, this is not always the case. *See e.g.*, U.S. Federal Energy Administration, *Energy Conservation in New Building Design: An Impact Assessment of ASHRAE Standard 90−75*, Conservation Paper No. 43B (Washington, D.C.: GPO, 1976), pp. 6−8.

53. For further discussion of this distinction, *see State and Local Government Purchasing, supra* note 8, pp. 11.6−11.7.

54. *See* General Services Administration (GSA) Order FSS 8020.1A, April 11, 1975.

55. GSA, *Value Incentive Clause*, GSA Form 2984 (October 1974) Sec. 1.1.

56. *Ibid.*, Sec. 8.

57. *Ibid.*, Secs. 8.1−8.4.

58. *Ibid.*, Sec. 6.

59. *See* 10 McQuillan, Municipal Corporations, § 29.101 (3rd ed. 1966).

60. *See e.g.*, *The Struggle to Bring Technology to the Cities* (Washington, D.C.: The Urban Institute, 1971); and National Conference of State Legislatures, *Meeting the Challenge* (Lexington, Kentucky: Council of State Governments, 1975).

61. *See State and Local Government Purchasing, supra* note 8, p. 6.6.

62. *Ibid.*, p. 6.5.

63. *Ibid.*, p. 11.6.

64. U.S. Environmental Protection Agency, *Third Report to Congress: Resource Recovery and Waste Reduction* (Washington, D.C.: GPO, 1975), pp. 79−80.

65. Note that the situation envisioned here, in contrast to the situation presented by the Washington and Florida laws in the life cycle costing subchapter, is one where design of a building system is carried out by the contractor which will ultimately construct it. The situation envisioned here may be preferable where the system is of an innovative nature and the contractor which will install it is much more familiar with it than the government's architects and engineers.

66. The asserted inappropriateness of state requirements is buttressed by the Federal Procurement Regulations, Title 41 C.F.R. In describing the application of the firm fixed price contract (essentially the type required by many state purchasing statutes), the regulations state that this contract type is "suitable for use in procurements when reasonably definite design or performance specifications are available. . . ." 41 C.F.R. Secs. 1−3.404−2(b).

67. Interested officials should contact: Procurement Programs, Experimental Technology Incentives Program, National Bureau of Standards, Washington, D.C. 20234.

Appendices to Chapter 1

THE PURCHASING PROCESS: A THUMBNAIL SKETCH

To clarify the intricacies of government purchasing for those interested in energy conservation and to share the perspective from which this study is undertaken, a brief, generalized discussion of the purchasing process follows.

Purchasing terminology is not difficult to master. For the purpose of this discussion, the purchasing process begins when the need for an item is first perceived by a potential user and ends when the contract award has been accomplished and goods or services in conformance with specifications have been received. "Potential users" are agencies, departments, or other governmental units of the jurisdiction in question.

Purchasing decisions made by potential users take into account need and budgetary constraints. Once these factors have had their play, it is time to prepare specifications for the needed item (or to choose among existing specifications) and to set about acquiring it.

A highly significant distinguishing characteristic among purchasing arrangements of the more than 78,000 state and local governments in the U.S. is the degree to which purchasing is centralized. If purchasing authority is centralized in a single agency, then that agency will acquire the supplies, services, or real estate for other government agencies. In such situations, the agency that needs an item, or the agency empowered to purchase it, or both, will prepare the specifications. All of the states have centralized purchasing systems [1]. Sixty four percent of the 1,169 cities responding to a

2,290 city survey of all cities with populations of 10,000 or more use central purchasing systems [2].

Of those governments with central purchasing, not all require that every acquisition be made by the purchasing agency. Exemptions are often made for designated agencies, for certain commodities, for purchases under a specified dollar amount, or for emergency purchases [3]. Thus, state and local purchasing is often handled by using agencies—either because of the complete lack of central purchasing mechanisms or because existing central purchasing requirements permit exemptions.

Central purchasing has several major benefits. Consolidating the needs of a number of using agencies for a particular item into a single contract increases the government's bargaining power by making the contract more attractive to potential suppliers. With this enhanced bargaining power, governments are more likely than individual agencies to elicit lower prices and better quality from contractors. Consolidating the purchasing function in a single agency is also apt to cut administrative costs. This savings, in turn, is likely to spur governments both to use more sophisticated purchasing techniques and to hire more experienced staff. If governments take up both challenges, they will compound their original savings on capital outlays and administrative costs.

In a well run central purchasing system, using agencies are required to estimate in advance their needs for a given period (usually six months or one year) and to place their orders with the centralized purchasing agency. This practice increases purchasing volume, and, thus, an agency's bargaining power.

Most purchasing agencies solicit competitive bids from potential suppliers. Exceptions to this procedure are similar to exceptions to centralized purchasing requirements [4]. Emergency purchases, necessary purchases from suppliers without competitors, and purchases below a certain dollar volume are usually made outside the competitive bidding system.

Where competitive bidding is employed, a list of qualified bidders is usually kept for each commodity. Qualification, if required, is based on capability to perform, and bidders may be removed from the list for failure to bid or for good cause (e.g., unsatisfactory performance or violation of standards).

Bids are solicited from all qualified bidders by means of an invitation for bids (IFB). The IFB contains the terms, conditions, and specifications that bidders rely on to prepare their proposals and that will ultimately be incorporated into the contract. When the purchase exceeds a certain dollar amount ($500 to $1,500 for states), most governments additionally require that legal notice of the solicitation be published in a widely circulated newspaper.

Where bidding is formal, bids are submitted in sealed envelopes that are then opened at a predetermined time and place that has been established in the IFBs [5]. The bid opening is generally open to the public.

Once opened, bids are evaluated comparatively. The following considerations may determine contract award:

1. Whether the bidder is "responsible." (This determination is based on the same factors that govern qualification for the bidders list, see p. 40. To some extent, prequalification for inclusion on the bidders list obviates this consideration).
2. Whether the bid is "responsive" (i.e., whether it complies with the specifications and contract terms as they are set out in the IFB).
3. Whether the bid price is the lowest among otherwise acceptable bids. (This consideration was discussed at length in Chapter 1's life cycle costing section beginning on p. 14.)
4. Whether there are other bid characteristics, such as quality or energy efficiency of the commodity in question, that are called for in the bid specifications to be included in the evaluation to determine the lowest responsive bidder.

Once the contract is awarded, the award is subject to challenge in the courts. The primary ground for challenge is that the award process did not comply with the purchasing statute of the jurisdiction in question—a problem that the author addresses both directly and indirectly in Chapters 1 through 3.

NOTES

1. *See* The Council of State Governments, *State and Local Government Purchasing* (Lexington, Kentucky: The Council of State Governments, 1975) Appendix A.
2. *Ibid.*, Appendix B.
3. *Ibid.*, pp. 3.2–3.3.
4. The Council of State Governments points out that state and local purchasing laws refer to "exceptions to," "exemptions from," and/or "waivers of" competitive bidding requirements. The council recommends use of the term "waiver" because this implies case-by-case determination, while "exceptions" and "exemptions" imply a blanket approach. Case-by-case determinations are preferred by the council because of frequently changing conditions and circumstances. *State and Local Government Purchasing*, p. 6.5. The term "waiver" is

not used in this discussion because emphasis is upon description rather than upon recommendation.

5. Florida Purchasing Regulations make the following distinction between formal and informal bidding: "(3) Formal Bid—A formal bid is defined as a sealed bid with the title, date and hour of the bid opening designated. (4) Informal Bid—An informal bid is defined as either a written or verbal quotation not requiring a public opening of such bid at a specific time or date." Florida General Regulations, Ch. 13A, §1.01, as amended May 19, 1972.

Informal competitive bidding sometimes fills the gap between purchases of a sufficiently low dollar volume not to require competitive bidding at all and those purchases greater than the amount set by statute to require formal bidding. For example, the state of Idaho requires informal bidding for purchases of between $100 and $1,000. *State and Local Government Purchasing*, p. A.5.

ENERGY EFFICIENCY STANDARDS

Proposed California Statutes

Governmental Acquisition of Motor Vehicles
The State of California shall not purchase or lease after September 1, 1976, any gasoline-powered motor vehicle having a manufacturer's gross weight rating of under 6,001 pounds and used primarily for the transportation of passengers, unless certified by its manufacturer as capable of operating at an overall miles-per-gallon ratio of 18 miles per one gallon of gasoline. [California Senate Bill No. 358, amending Chapter 21, introduced February 5, 1975.]

Limitations on Passenger Vehicle Acquisition
Passenger vehicle acquisitions of a public agency for each year shall be such that no more than 10 percent of such vehicles have a fuel economy of less than 16 miles per gallon for the year 1976, 18 miles per gallon for the year 1977, and 20 miles per gallon for the year 1978 and each subsequent year thereafter. [California Assembly Bill No. 758, introduced February 10, 1975.]

State Purchases of Passenger Vehicles
Except for patrol cars of the Department of the California Highway Patrol, the state shall not purchase, lease, or otherwise acquire passenger vehicles, as defined by Section 465 of the Vehicle Code, that have a fuel economy of less than 17 miles per gallon as certified by an average of simulated city driving and simulated highway driving of the most recent test results with respect to fuel economy for passenger vehicles sold in California of the most recent tests conducted by the Federal Environmental Protection Agency pursuant to Section 206(e) of the Clean Air Act, as amended (42 U.S.C. 1856 f-5(e)). [California Assembly Bill No. 963, introduced February 27, 1975.]

LIFE CYCLE COSTING   ˙

State of Florida
Energy Conservation in Buildings Act

§255.251—Short Title
This act shall be cited as the "Florida Energy Conservation in Buildings Act of 1974."

§255.252—Findings and Intent
(1) Operating and maintenance expenditures associated with energy equipment and with energy consumed in state-financed and leased buildings represent a significant cost over the life of a building. Energy conserved by appropriate building design not only reduces the demand for energy but also reduces costs for building operation. For example, commercial buildings are estimated to use from 20 to 80 percent more energy than would be required if energy-conserving designs were used. The size, design, orientation, and operability of windows, the ratio of ventilating air to air heated or cooled, the level of lighting consonant with space-use requirements, the handling of occupancy loads, and the ability to zone off areas not requiring equivalent levels of heating or cooling are but a few of the considerations necessary to conserving energy.

(2) Significant efforts are underway by the General Services Administration, the National Bureau of Standards, and others to detail the considerations and practices for energy conservation in buildings. Most important is that energy-efficient designs provide energy savings over the life of the building structure. Conversely, energy-inefficient designs cause excess and wasteful energy use and high costs over that life. With buildings lasting many decades and with energy costs escalating rapidly, it is essential that the costs of operation and maintenance for energy-using equipment be included in all design proposals for state buildings.

(3) In order that such energy efficiency considerations become a function of building design, and also a model for future application in the private sector, it shall be the policy of the state that buildings constructed and financed by the state be designed and constructed in a manner which will minimize the consumption of energy used in the operation and maintenance of such buildings.

§255.253—Definitions
(1) "Division" means the Division of Building Construction and Maintenance of the Department of General Services.

(2) "Facility" means a building or other structure.

(3) "Energy performance index or indices" (EPI) means a number of describing the energy requirements at the building boundary of a facility, per square foot of floor space or per cubic foot of occupied volume, as appropriate under defined internal and external ambient conditions over an entire seasonal cycle. As experience develops on the energy performance achieved with state building, the indices (EPI) will serve as a measure of building performance with respect to energy consumption.

(4) "Life-cycle costs" means the cost of owning, operating, and maintaining the facility over the life of the structure. This may be expressed as an annual cost for each year of the facility's use.

§ 255.254—No Facility Constructed or Leased Without Life-Cycle Costs

No state agency shall lease, construct, or have constructed, within limits prescribed herein, a facility without having secured from the division a proper evaluation of life-cycle costs, as computed by a qualified architect or engineer. Furthermore, construction shall proceed only upon disclosing, for the facility chosen, the life-cycle costs as determined in § 255.255 and the capitalization of the initial construction costs of the building. The life-cycle costs shall be a primary consideration in the selection of a building design. Such analysis shall be required only for construction of buildings with an area of 5,000 square feet or greater. For leased areas of 20,000 square feet or greater within a given building boundary, a life-cycle analysis shall be performed and a lease shall only be made where there is a showing that the life-cycle costs are minimal compared to available like facilities.

§ 255.255—Life-Cycle Costs

(1) The division shall promulgate rules and procedures, including energy conservation performance guidelines, for conducting a life-cycle cost analysis of alternative architectural and engineering designs and for developing energy performance indices to evaluate the efficiency of energy utilization for competing designs in the construction of state-financed and leased facilities. Such rules and procedures shall take effect 270 days after the enactment of this law.

(2) Such life-cycle costs shall be the sum of:

(a) The reasonably expected fuel costs over the life of the building, as determined by the division, that are required to maintain illumination, power, temperature, humidity, and ventilation and all other energy-consuming equipment in a facility, and

(b) The reasonable costs of probable maintenance, including labor and materials, and operation of the building.

(3) To determine the life-cycle costs as defined in subsection (2)(b), the department shall promulgate rules that shall include, but not be limited to:

(a) The orientation and integration of the facility with respect to its physical site.

(b) The amount and type of glass employed in the facility and the directions of exposure.

(c) The effect of insulation incorporated into the facility design and the effect on solar utilization of the properties of external surfaces.

(d) The variable occupancy and operating conditions of the facility and subportions of the facility.

(e) An energy consumption analysis of the major equipment of the facility's heating, ventilating, and cooling system, lighting system, hot water system, and all other major energy-consuming equipment and systems as appropriate. This analysis shall include:

1. The comparison of alternative systems.
2. A projection of the annual energy consumption of major energy-consuming equipment and systems for a range of operations of the facility over the life of the facility.
3. The evaluation of the energy consumption of component equipment in each system, considering the operation of such components at other than full or rated outputs.

(4) Such rules shall be based on the best currently available methods of analysis, including such as those of the National Bureau of Standards, the Department of Housing and Urban Development, and other federal agencies and professional societies and materials developed by the department. Provisions shall be made for an annual updating of rules and standards as required.

§ 255.256—Energy Performance Index

The department shall promulgate rules for energy performance indices as defined in § 255.253(3) to audit and evaluate competing design proposals submitted to the state.

State of Washington
Energy Conservation in
Design of Public Facilities

§ 39.35.010—Legislative Findings

The legislature hereby finds:

(1) That major publicly owned or leased facilities have a significant impact on our state's consumption of energy;

(2) That energy conservation practices adopted for the design, construction, and utilization of such facilities will have a beneficial effect on our overall supply of energy;

(3) That the cost of energy consumed by such facilities over the life of the facilities shall be considered in addition to the initial cost of constructing facilities; and

(4) That the cost of energy is significant and major facility designs shall be based on the total life-cycle cost, including the initial construction cost, and the cost, over the economic life of a major facility, of the energy consumed, and the operation and maintenance of a major facility as they affect energy consumption. [Added by Laws 1st Ex. Sess. 1975 Ch. 177 §1.]

§ 39.35.020—Legislative Declaration

The legislature declares that it is the public policy of this state to insure that energy conservation practices are employed in the design of major publicly owned or leased facilities. To this end the legislature authorizes and directs that public agencies analyze the cost of energy consumption of each major facility to be planned and constructed or renovated after September 8, 1975. [Added by Laws 1st Ex. Sess. 1975 Ch. 177 §2.]

§39.35.030 — Definitions

For the purposes of this chapter the following words and phrases shall have the following meanings unless the context clearly requires otherwise:

(1) "Public agency" means every state office, officer, board, commission, committee, bureau, department, and all political subdivisions of the state.

(2) "Major facility" means any publicly owned or leased building having twenty-five thousand square feet or more of usable floor space.

(3) "Initial cost" means the moneys required for the capital construction or renovation of a major facility.

(4) "Renovation" means additions, alterations, or repairs within any twelve month period which exceed fifty percent of the value of a major facility and which will affect any energy system.

(5) "Economic life" means the projected or anticipated useful life of a major facility as expressed by a term of years.

(6) "Life-cycle cost" means the cost of a major facility including its initial cost, the cost of the energy consumed over its economic life, and the energy consumption related cost of its operation and maintenance.

(7) "Life-cycle cost analysis" includes, but is not limited to, the following elements:

(a) The coordination and positioning of a major facility on its physical site;

(b) The amount and type of fenestration employed in a major facility;

(c) The amount of insulation incorporated into the design of a major facility;

(d) The variable occupancy and operating conditions of a major facility; and

(e) An energy-consumption analysis of a major facility.

(8) "Energy systems" means all utilities, including, but not limited to, heating, air-conditioning, ventilating, lighting, and the supplying of domestic hot water.

(9) "Energy-consumption analysis" means the evaluation of all energy systems and components by demand and type of energy including the internal energy load imposed on a major facility by its occupants, equipment, and components, and the external energy load imposed on a major facility by the climatic conditions of its location. An energy consumption analysis of the operation of energy systems of a major facility shall include, but not be limited to, the following elements:

(a) The comparison of three or more system alternatives;

(b) The simulation of each system over the entire range of operation of such facility for a year's operating period; and

(c) The evaluation of energy consumption of component equipment in each system considering the operation of such components at other than full or rated outputs.

The energy-consumption analysis shall be prepared by a professional engineer or licensed architect who may use computers or such other methods as are capable of producing predictable results. [Added by Laws 1st Ex. Sess. 1975 Ch. 177 §3.]

§ 39.35.040—Facility Design to
Include Life-Cycle Cost Analysis

On and after September 8, 1975 whenever a public agency determines that any major facility is to be constructed or renovated such agency shall cause to be included in the design phase of such construction or renovation a provision that requires a life-cycle cost analysis to be prepared for such facility. Such analysis shall be approved by the agency prior to the commencement of actual construction or renovation. A public agency may accept the facility design if the agency is satisfied that the life-cycle cost analysis provides for an efficient energy system or systems based on the economic life of the major facility. [Added by Laws 1st Ex. Sess. 1975 Ch. 177 §4.]

§ 39.35.900—Severability—1975 1st Ex. Sess. Ch. 177

If any provision of this act, or its application to any person or circumstance is held invalid, the remainder of the act, or the application of the provision to other persons or circumstances is not affected. [Added by Laws 1st Ex. Sess. 1975 Ch. 177 §5.]

U.S. General Services Administration
Energy Conservation Design Guidelines:
Life Cycle Cost Analysis

Life cycle cost analysis is a method of trade-off among sub-systems primarily from a cost viewpoint. Sub-systems are of two general categories:

(1) Those that involve little or no maintenance or operating costs, such as double glazing; and,

(2) Those that require labor and materials for maintenance and operations, such as boilers.

With raw energy use as a primary consideration, life cycle analysis should include:

(1) Selection of alternative systems for analysis.

(2) Designation of one system as the base line system and ascertaining the initial incremental cost differential among each of the systems under analysis. Where selection of any one sub-system will affect the initial cost of another system, the net change in initial costs of the sub-system and others it affects must be considered. For example, the use of insulation may reduce the size and capital cost of the heating and/or cooling system.

(3) The number of years of useful life of the total building* being estab-

---

*Use 40 years life for must buildings. Justify a lesser amount by analysis of trends, type of construction, intended use. Use caution when being advised of

lished prior to determining the differential owning costs (interest and amortization of the sub-systems). The useful life of each sub-system or each portion of the sub-system must be assumed, given normal maintenance procedures. Various portions of sub-systems may have different life expectancies. For instance, in central air-conditioning systems, the duct work will normally outlast the compressor and cooling tower.

(4) Neglecting the residual value of the sub-system at the end of the useful life of the building in most cases. However, in a building with a relatively short life expectancy, such as a temporary building or an inflatable structure, the sub-system may have a re-sale value. Calculations for amortization must take that into account. Beware, however, of claiming a residual value for a piece of equipment which may have portions perfectly intact, although other portions may have outlived their usefulness. For example, the casing of a window air conditioner may be in perfect condition after the coils have corroded beyond repair. In such cases, the useful life of the initial unit coincides with the life of the coils. Likewise, insulation may be in good condition at the end of the useful life of the building, but cannot be salvageable because of prohibitive cost to remove, or inaccessibility.

(5) Selection of a rate of interest which is generally applied equally to the differential cost of each sub-system. When interest rates are fluctuating and it is not possible to establish a current stable rate, a rate of ½ of 1% over the prime rate may be used. Various organizations, government, and owners enjoy different borrowing rates. In some cases, interest rates may vary among sub-systems. If so, an individual rate must be applied to each incremental cost difference. Some higher risk or innovative systems may not be able to be financed at rates prevailing at the time of construction.

(6) The differential annual owning costs which can now be determined by multiplying the capital recovery factor, CRF, by each individual incremental cost differential between its respective system and the base-line system. The capital recovery factor may be found in such references as Grant's *Engineering Economics*. Tables include the CRF for various interest rates and period of amortization (the expected useful life of the system).

Where the life expectancy of the sub-system is truly realizable, the incremental initial cost differential must be included for that sub-system which will need to be replaced in full prior to the demise of the building. The replacement value must be based on the estimated cost at the time of replacement, reflecting the anticipated rate of inflation. Interest rates at the time of replacement are unknown now. Applying current rates equally to all sub-systems is satisfactory for this analysis.

Life cycle costs include (in addition to the sum of annual interest and amortization of the incremental differences in initial costs) the sum of annual operating costs which consist of fuel and electric power costs, maintenance and operation materials, and labor costs. When only differential life cycle costs are of interest,

---

short anticipated life (many "temporary World War II buildings are still in use) and remember that it is the building life, not the owners' (Government) expected duration of occupancy that governs energy conservation.

the differentials among sub-systems rather than absolute values are required for maintenance and operation. Normal annual increases in labor, materials, and energy costs can be expected. At present, energy costs are escalating more rapidly than labor and materials. Energy costs may double within five years—so the calculations should assume that they will—and then perhaps a lesser percent each year thereafter. Labor and materials should be calculated as increasing each year.

The annual owning and operating costs differential between the base line system and each alternative system can be compared, the net difference being the algebraic sum of the owning and operating costs of each system. Multiplying the assumed life expectancy of the building by the annual differential owning and operating costs yields the differential life cycle costs of each system.

If absolute life cycle costs are required for any one or more systems, the true initial costs and the sum of the gross annual owning and operating costs must be determined rather than the differential sums among the systems.

Cost benefit analyses are useful in examining the relative merits of comparative systems in relation to their initial or life cycle costs. These benefits may be realized in comfort, improved worker efficiency, acoustical control, reduced air or water pollution, conservation of natural, non-renewable resources, a more pleasing, aesthetic environment, reduced need for community services, or a multitude of other considerations, none of which are easily dollar identifiable.

[Public Building Service, U.S. General Services Administration, *Energy Conservation Design Guidelines for New Office Buildings*, 2d ed. (Washington, July 1975), Appendix D.]

GSA Value Incentive Clause
[Fixed Price Supply Contract]

1. Intent and Objectives
This clause applies to any cost reduction proposal (hereinafter referred to as a Value Change Proposal or VCP) submitted after award by the Contractor for the purpose of changing any requirement of this contract. VCP's may be submitted only by Contractors holding current contracts and only for those items awarded to the Contractor. This clause does not, however, apply to any such proposal unless it is identified by the Contractor, at the time of its submission to the Contracting Officer, as a proposal submitted pursuant to this clause.

1.1 VCP's contemplated are those that would result in net savings to the Government by providing either: (1) a decrease in the cost of performance of this contract, or; (2) a reduction in the cost of ownership (hereinafter referred to as collateral costs as determined in accordance with paragraph 6) of the work provided by this contract, regardless of acquisition costs. VCP's must result in savings without impairing any required functions and characteristics such as service life, reliability, economy of operation, level of operational performance, ease of maintenance, standardized features, esthetics, fire protection features, and safety features, presently required by this contract. However, nothing herein precludes the submittal of VCP's where the Contractor considers that the required functions and characteristics could be combined, reduced, or eliminated, as being non-essential or excessive to the satisfactory performance of the work involved.

1.2 A VCP identical to one submitted under any other contract with the Contractor may also be submitted under this contract.

2. Subcontractor Inclusion

The Contractor shall include the provisions of this clause, with a provision for sharing arrangements that meet or exceed the minimum percentage contained in paragraph 8.2, in all subcontracts in excess of $25,000, and, in any other subcontract which, in the judgment of the Contractor, is of such nature as to offer reasonable likelihood of value change proposals. At the option of each Subcontractor, this clause may be included in lower tier subcontracts. The Contractor shall encourage submission of VCP's from Subcontractors; however, it is not mandatory that VCP's be submitted nor is it mandatory that the Contractor accept and/or transmit to the Government VCP's proposed by his Subcontractors.

3. Data Requirements

As a minimum, the following information shall be submitted by the Contractor with each VCP:

3.1 A description of the difference between the existing contract requirement and the proposed change, and the comparative advantages and disadvantages of each; including justification where function or characteristic of the work is being reduced;

3.2 Separate detailed cost estimates for both the existing contract requirement and the proposed change, and an estimate of the change in contract price including consideration of the costs of development and implementation of the VCP and the sharing arrangement set forth in the clause;

3.3 An estimate of the effects the VCP would have on collateral costs (as determined in accordance with paragraph 6) to the Government, including an estimate of the sharing that the Contractor requests be paid by the Government upon the approval of the VCP;

3.4 Technical analysis in sufficient detail to identify and describe each requirement of the contract which must be changed if the VCP is accepted, with recommendation as to how to accomplish each such change, and its effect on unchanged work;

3.5 A statement of the time by which approval of the VCP must be issued by the Government to obtain the maximum cost reduction, noting any effect on the contract completion time or delivery schedule; and

3.6 Identification of any previous submission of the VCP, including the dates submitted, the agencies involved, the numbers of the Government contracts involved, and the previous actions by the Government if known.

4. Processing Procedures

Six copies of each VCP shall be submitted to the Contracting Officer. VCP's will be processed expeditiously; however, the Government will not be liable for any delay in acting upon a VCP submitted pursuant to this clause. The Contractor may withdraw, in whole or in part, a VCP not accepted by the Government within the period specified in the VCP. The Government shall not be liable for

VCP development cost in the case where a VCP is rejected or withdrawn. The decision of the Contracting Officer as to the acceptance or rejection of a VCP under this contract shall be final and shall not be subject to the "Disputes" clause of this contract.

4.1 The Contracting Officer may modify a VCP, with concurrence of the Contractor, to make it acceptable, and the Contractor's share will be based on the VCP as modified.

4.2 Pending written acceptance of a VCP in whole or in part, the Contractor shall remain obligated to perform in accordance with the terms of the existing contract.

4.3 An approved VCP shall become effective through written notice and/or issuance of an equitable adjustment in the contract price and/or time of performance by a contract modification pursuant to the provisions of this clause bearing a notation so stating. Where an approved VCP also involves any other applicable clause of this contract such as "Termination for Convenience of the Government," "Suspension of Work," or "Changes," then that clause shall be cited in addition to this clause.

5. Computations for Change in Cost of Performance

The Contractor shall prepare separate estimates for both the existing (Instant) contract requirement and the proposed change. Each estimate shall consist of an itemized breakdown of all costs of the Contractor and all Subcontractors' work in sufficient detail to show unit quantities and total costs of labor, material, overhead and equipment.

5.1 Contractor development and implementation costs for the VCP shall be included in the estimate for the proposed change. However, these costs will not be allowable if they are otherwise reimbursable as a direct charge under this contract.

5.2 If the VCP results in a net reduction in contract price, Government costs of processing or implementation of a VCP shall not be included in the estimate.

5.3 If the difference in the estimates indicate a net reduction in contract price, the profit of the original contract price will remain the same and will not change as a result of the VCP. The resultant net reduction in contract cost of performance shall be shared as stipulated hereinafter.

5.4 If the difference in the estimates, as approved by the Government, indicates a net increase or decrease in contract price, the price shall be adjusted by contract modification.

6. Computations for Collateral Costs

The Contractor shall prepare separate estimates for collateral costs of both the existing requirement and the proposed change. Each estimate shall consist of an itemized breakdown of all costs and the basis for the data used in the estimate. Cost benefits to the Government include, but are not limited to: reduced costs of operation, maintenance or repair, extended useful service life, replacement or future acquisition, logistical support, and reduction in requirements for Government furnished property. Increased collateral costs include the converse of such factors. Computation shall be as follows:

6.1 Costs shall be calculated over a 10-year period on a uniform basis for each estimate and shall include the Contractor's best estimate of the Government's costs of processing or implementing the VCP.

6.2 If the difference in estimates indicates a savings, the Contractor shall divide the resultant amount by 10 to arrive at the average annual net collateral savings. The resultant savings shall be shared as stipulated hereinafter.

6.3 The determination of the reasonableness, allocability and allowability of collateral costs contained in the Contractor's estimates shall be made by the Contracting Officer. In the event that agreement cannot be reached on the amount of the estimated collateral costs the Contracting Officer shall determine the amount. In either case, the Contracting Officer's decision will be final and not subject to the provisions of the "Disputes" clause of this contract.

7. Computations for Future Acquisitions

If a VCP is accepted under this clause and used in future solicitations the Contractor will be paid a royalty share of savings realized by the General Services Administration on future purchases, if any, of items utilizing the VCP or on modifications made to other existing contracts to utilize the VCP within the royalty sharing period.

7.1 To qualify for royalty sharing, the VCP must result in a unit price reduction for an item under this contract. The Contractor will be paid for the actual number of items purchased during, and limited to, the periods of time as follows; for indefinite quantity and requirements contracts, the next 12 month contract period following the expiration of this contract; or for definite quantity contracts, any solicitation issued that utilizes the VCP during the next 12 month period commencing with the date of modification of this contract accepting the VCP.

7.2 For purposes of determining the Contractor's royalty share, the "unit price reduction" under this contract is the Contracting Officer's estimate of the effect which the VCP would have had on the Contractor's cost of performance if the change had been included in the original specifications under this contract (this estimate shall not take into account any costs of developing or implementing the change).

8. Sharing Arrangements

If a VCP is accepted by the Government, the Contractor is entitled to share in instant and/or future contract savings, or collateral savings to the full extent provided for in this clause. For purposes of sharing under this clause, the term "instant contract" shall not include any changes to or other modifications of this contract, executed subsequent to acceptance of the particular VCP, by which the Government increases the quantity of any item or work or adds any item of work. It shall not include any extension of the instant contract through exercise of an option (if any) provided under this contract after acceptance of the VCP. Such actions shall be eligible for future acquisition savings.

If a VCP accepted by the Government results in a net reduction in contract price, the Contractor is entitled to share in instant and future acquisition savings but not in collateral savings. If a VCP, accepted by the Government, results in a net increase in contract price, the Contractor shall share in collateral savings, but not in future acquisition savings. In case of accepted price increases, the current contract will be modified to allow for the increase in price plus the amount allowed for collateral savings for the remainder of the curent (instant) contract period. The amount of sharing shall be at the rates provided below.

If the Contractor submits under this clause a proposal which is identical, or substantially similar, to one previously received by the Contracting Officer under a different contract with the Contractor for substantially the same terms and both proposals are accepted by the Government, the Contractor shall share instant contract savings realized under this contract, pursuant to paragraph 8.1 of this clause, but he shall not share future acquisition savings.

8.1 If the prime Contractor is solely responsible for the VCP, he shall receive 50% and the Government 50% of the net reduction in cost of performance of this contract.

8.2 If a Subcontractor is responsible for the VCP, the prime Contractor agrees that the Subcontractor shall receive a minimum of 25%, the prime Contractor a maximum of 25%, and the Government a fixed 50%, of the net reduction in the cost of performance of this contract. Other Subcontractors shall receive a portion of the first-tier Subcontractor savings in accordance with the terms of their contract with first-tier Subcontractor.

8.3 When collateral savings occur the Contractor shall receive 20% of the average one year's net collateral savings.

8.4 When future acquisition savings occur, the Contractor shall receive a royalty share equal to 30% of the unit cost reduction realized under the instant contract.

9. Adjustment to Contract Price

9.1 The method for payment of instant savings shares shall be accomplished by reducing the contract unit price by an amount equal to the Government's share of the savings.

9.2 Payments for accepted VCP's involving price increases and collateral sharing will be accomplished by adjusting the current contract unit price by the amount of the increase plus the increase as provided for in paragraph 8.3. All orders placed after acceptance of the VCP will be paid for at the increased unit price until expiration of the current (Instant) contract.

9.3 Payments for accepted VCP's involving price decreases and royalty sharing will be accomplished by adjusting the current contract unit price by the price decreases and, if the current Contractor submitting the accepted VCP receives the "future" contract, that unit price will be increased, as provided for in paragraph 8.4, at the time of award for the new contract period.

If a different Contractor receives the future contract, the Contractor submitting the accepted VCP shall be paid, as provided for in paragraph 8.4, by multiplying the savings by the number of units actually purchased by the Government. In this case, royalty shares will be permitted to accumulate and paid quarterly.

10. Data Restriction Rights

The Contractor may restrict the Government's right to use any sheet of a VCP or of the supporting data, submitted pursuant to this clause, in accordance with the terms of the following legend if it is marked on each such sheet:

> The data furnished pursuant to the Value Incentive Clause of contract* shall not be disclosed outside the Government, or duplicated, used, or disclosed in whole or in part, for any purpose other than to evaluate a VCP submitted under said clause. This restriction does not limit the Government's right to use information contained in this data if it is or has been obtained or is otherwise available, from the Contractor or from another source, without limitations, nor, shall this restriction apply in any respect after a period of two years from the date the VCP is submitted. If such a proposal is accepted by the Government under said contract after the use of this data in such an evaluation the Government shall have the right to duplicate, use, and disclose any data reasonably necessary to the full utilization of such proposal as accepted, in any manner and for any purpose whatsoever, and have others so do.

In the event of acceptance of a VCP the Contractor hereby grants to the Government all rights to use, duplicate or disclose, in whole or in part, in any manner and for any purpose whatsoever, and to have or to permit others to do so, data reasonably necessary to fully utilize such proposal on this and any other Government contract.

In lieu of repeating this above legend, Contractors may use a reference as follows on the appropriate sheet, "This sheet is restricted as provided in paragraph 10 of the Value Incentive Clause of Contract number.* "

[U.S. General Services Administration, GSA Form 2984 (Oct. 1974).]

VCP ENERGY SAVINGS

The FSS Value Incentive Clause (VIC) under the Value Management Program has already proven productive since its inception during June 1975. Two VCPs have been received involving energy reductions.

The A.C. Manufacturing Company, Cherry Hill, New Jersey, has submitted a VCP dated October 2, 1975, under FSS Region 3 Contract No. GS−03S−44536 for FSC 4120 covering twenty 7.5 ton capacity Computer Room Type Air Conditioners in the total amount of $68,900. The VCP has been accepted and the contract modified reflecting the following:

● Lower installation cost. The current draw of 96 amps at 220 V per unit is reduced to 46.1 amps resulting in a unit savings of 17.6 KVA ($968 per unit).

---

*Contractor will insert the applicable contract number.

- Lower electrical demand charge. The reduction in the current draw results in a savings of 16.7 KW per unit a month ($33.40 per unit a month).
- Lower electrical operating cost. A total 16.7 KW savings per unit a month results in 876 hours (10 percent of year) for dehumidification cycle. This computes to 876 hours × 16.7 KW = 1463 KWH per unit per year ($658 per unit per year).
- Substitution of panel type humidifier using hot gas water heater in place of pan type humidifier using electric heat. Humidifiers operate 30 percent of 8760 hours or 2628 hours × 3.45 KW per unit = 9053 KWH per year savings (@ $.045 per KWH = $407 per unit a year). The panel type humidifier also results in similar 9053 KWH savings per unit a year for the cooling cycle. Total savings computes to 18,106 KWH per unit a year ($814 per unit a year).

The estimated savings and item increases are as follows:

Total ten year savings

Installation cost savings	$ 19,360
Reduced electrical demand	80,160
Reduced electrical operating cost	131,663
Humidifier operating cost	162,962
Total savings	$394,145
Contractor's share $\dfrac{394{,}145}{10}$ × 20%	7,883
Increase in unit cost	7,650
Total increase in contract price	$ 15,333

Liebert Corporation, Columbus Ohio, has submitted a VCP dated November 11, 1975, under FSS Region 3 Contract No. GS–03S–4437 for FSC 4120 covering twenty 10 ton and twenty 15 ton computer room type air conditioners in the total amount of $182,875. This VCP has been approved covering the following changes:

- Provide semi-hermetically sealed compressors in place of hermetically sealed compressors which require 22 percent more energy.
- Provide two 5 ton compressors on the 10 ton units in place of one 10 ton compressor, resulting in 50 percent less energy.
- Provide solid state temperature and humidity controls with design life of fifteen years and reduction in maintenance costs.

The estimated savings and item increases are as follows:

Total ten year energy cost savings $378,511

Contractor's share $\dfrac{378{,}511}{10}$ × 20% =   7,570

Increase in cost of items

10 ton units	30,354
15 ton units	28,165
Total increase in contract price	$66,089

Source: Federal Supply Service, Draft Memorandum, December 23, 1975, courtesy of Mr. Laird Smith.

AUTHOR'S SUGGESTED LEGISLATIVE APPROACH #1

IMPORTANT NOTICE: This draft legislative approach constitutes no more than suggestions with respect to the problems posed. It should, therefore, be introduced only after careful consideration of local conditions. Existing constitutional and statutory requirements should be thoroughly examined. Revisions in the statutory language, section headings, numberings, and other modifications may be necessary in order to conform to local law and practices.

Section 1. This act shall be called the "(state or municipal) Motor Vehicle Fuel Efficiency and Energy Conservation Act of 197_ ."

Section 2. There is a substantial present need to conserve energy and to improve the economy of government operations. Procurement of fuel-efficient passenger motor vehicles for (state or municipal) use will further these goals within the context of (state or municipal) operations and, further, will encourage the conservation of energy in the private sector. Private sector conservation will be encouraged because (1) purchase of fuel-efficient passenger motor vehicles with greater fuel efficiency and (2) the (state's or municipality's) experience with use of fuel-efficient passenger motor vehicles will serve as a model for private enterprise and individual consumers.

Section 3. For the purpose of this act:
(1) The term "passenger motor vehicle" means a gasoline-propelled motor vehicle, designed for carrying nine persons or less, except (A) a motorcycle or other similar vehicle, (B) a truck not designed primarily to carry its operator or passengers, or (C) a police, fire, or other emergency vehicle.
(2) The term "Director" means the Director of the (Department of General Services or other similar agency).
(3) The term "fleet average fuel efficiency" means (A) the total number of passenger motor vehicles acquired in a fiscal year by all public agencies divided by (B) a sum of terms, each term of the sum being a fraction created by dividing

(1) the number of passenger motor vehicles of a given model type acquired by public agencies by (2) the fuel efficiency of that model type.

(4) The term "acquire" means to lease for a period of (thirty) days or more, or to purchase.

(5) The term "public agency" includes an office, department, bureau, board, commission, or agency of the (state or municipality).

Section 4. (a) The Director shall promulgate within _____ days of the effective date of this act regulations ensuring that the fleet average fuel efficiency of all passenger motor vehicles acquired during the next fiscal year and subsequent fiscal years by all public agencies shall be at least _____ miles per gallon greater than the average fuel economy standard applicable under §502(a) of the Motor Vehicle Information and Cost Savings Act, 15 U.S.C. §2002(a). The federal standard for the model year that includes the first day of the state fiscal year shall be the standard applicable to that fiscal year.

(b) Records of fuel efficiency shall be kept for each passenger motor vehicle acquired under the provisions of this act. Records for individual vehicles shall be compiled at the end of each fiscal year and submitted to the director.

(c) Within _____ days after the end of each fiscal year, the director shall submit to the (legislature or local lawmaking body) an annual report based upon the record keeping and setting out:

(1) the number of each model type and the fuel efficiency of each model type, according to the U.S. Environmental Protection Agency ratings, that was acquired during that fiscal year;

(2) the actual average fuel efficiency of each model type and of all passenger motor vehicles acquired during that fiscal year according to the records kept pursuant to subsection 4(b) of this act;

(3) an assessment of the effect of this acquisition policy upon the effectiveness and the cost of government operations; and

(4) a recommendation of whether the required fleet average fuel efficiency standard should be made more or less stringent.

Section 5. If any provision of this act or its applicability to any person or circumstance is held invalid, the remainder of the act, or the application of the provision to other persons or circumstances, is not affected.

AUTHOR'S SUGGESTED LEGISLATIVE APPROACH #2

IMPORTANCE NOTICE: This draft legislative approach constitutes no more than suggestions with respect to the problems posed. It should, therefore, be introduced only after careful consideration of local conditions. Existing constitutional and statutory requirements should be thoroughly examined. Revisions in the statutory language, section headings, numberings, and other modifications may be necessary in order to conform to local law and practices.

Sections 1 and 2. [Same as Sections 1 and 2 of Author's Proposed Legislative Approach #1.]

Section 3. For the purpose of this act;
§ (1), (2), and (3) [Same as parallel subsections in Author's Proposed Legislative Approach #1.]
(4) The term "rank" means a point on a listing of models of motor vehicles rated for fuel efficiency by the United States Environmental Protection Agency (hereinafter, EPA), where the listing is in order of fuel efficiency with the most fuel-efficient model occupying the top rank and the least fuel-efficient model occupying the lowest rank. (EPA ratings are issued as an incident to emission testing conducted by the EPA pursuant to the federal Clean Air Act, 42 U.S.C. §§ 1857–1858a (1970).)
§ (5) and (6) [Same as (4) and (5), respectively, in Author's Proposed Legislative Approach #1.]

Section 4. (a) The director shall promulgate within _____ days of the effective date of this act regulations ensuring that the fleet average fuel efficiency of all passenger motor vehicles acquired during the next full fiscal year and subsequent fiscal years by all public agencies shall be equal to or greater than the fuel efficiency rating of the model of motor vehicle that occupies a rank that is __(5)__ models above the middle of the listing. The latest set of EPA ratings issued prior to the beginning of a fiscal year shall be listed to determine the required rank for that fiscal year.
(b) and (c) [Same as parallel subsections of Author's Proposed Legislative Approach #1.]

Section 5. [Severability Clause, as in Author's Proposed Legislative Approach #1.]

AUTHOR'S SUGGESTED LEGISLATIVE APPROACH #3

IMPORTANT NOTICE: This draft legislative approach constitutes no more than suggestions with respect to the problems posed. It should, therefore, be introduced only after careful consideration of local conditions. Existing constitutional and statutory requirements should be thoroughly examined. Revisions in the statutory language, section headings, numberings, and other modifications may be necessary in order to conform to local law and practices.

Section 1. This act shall be called "The Modified Life Cycle Costing Act of 197_ ."

Section 2. [Preamble. (Need to save energy and money in government operations.)]

Section 3. For the purposes of this act:
(1) The term "energy-consuming commodities" means products that consume energy to perform their primary function. Specifically, these products

include motor vehicles of any type, refrigerators, refrigerator-freezers, freezers, room air conditioners, central air conditioners, space heating equipment, humidifiers, dehumidifiers, and any other product so prescribed by the Director.

(2) The term "modified life cycle cost" of an energy-consuming commodity means the price to be paid by the government to acquire that commodity plus the projected cost of energy required to operate that commodity for a period of time that approximates the projected useful life of that commodity.

(3) The term "Director" means the Director of the (Department of General Services or other similar agency).

(4) The term "public agency" includes an office, department, bureau, board, commission, or agency of the (state or municipality).

Section 4. (a) All bids submitted for contracts to supply energy-consuming commodities to any public agency shall be evaluated according to the modified life cycle cost of the commodities. Contract award shall be made to the responsible bidder who submits a responsive bid with the lowest modified life cycle cost.

(b) The following information shall be determined by the Director for each type of energy-consuming commodity to be acquired by any public agency, and this information shall be set out clearly in invitations to bid for contracts to supply energy-consuming commodities:

(1) The number of years of useful life and the number of hours of annual operation, if applicable, upon which the calculation of the cost of energy required to operate such commodity shall be based;

(2) The price of energy upon which the calculation of energy cost shall be based;

(3) The testing method by which the average hourly energy consumption or other measure of energy consumption of the commodity is to be ascertained; and

(4) The mathematical formula for ascertaining the modified life cycle cost of the commodity.

Section 5. [Severability.]

AUTHOR'S SUGGESTED LEGISLATIVE APPROACH #4

IMPORTANT NOTICE: This draft legislative approach constitutes no more than suggestions with respect to the problems posed. It should, therefore, be introduced only after careful consideration of local conditions. Existing constitutional and statutory requirements should be thoroughly examined. Revisions in the statutory language, section headings, numberings, and other modifications may be necessary in order to conform to local law and practices.

Section 1. This act shall be called the "Life Cycle Costing for Government Purchase of Energy-Consuming Commodities Act of 197_ ."

Section 2. [Preamble.]

Section 3. This act applies to government contracts of greater than, or anticipated to be greater than, $ _____ for the procurement of energy-consuming commodities.

Section 4. For the purposes of this act:

(1) The term "energy-consuming commodities" means products that consume energy to perform their primary functions. Specifically, these include motor vehicles of any type, refrigerators, refrigerator-freezers, freezers, room air conditioners, central air conditioners, space heating equipment, humidifiers, dehumidifiers, and any other product so prescribed by the <u>Director.</u>

(2) The term "life cycle cost" of a commodity means the sum total of financial costs that are incurred as a result of acquiring that commodity. These costs may include purchase price and other acquisition costs, operating and maintenance costs, costs of modifying government operations to permit the introduction of the product, and costs of disposing of the product. Disposition costs may be negative if the product has a salvage value at the end of its projected period of use. All future costs are discounted to present value at a rate set by the <u>Director.</u>

(3) The term "average yearly cost of ownership" means the life cycle cost divided by the anticipated period of product use. Where ascertainable and readily available, this anticipated period may vary among competing products, according to their durability. Where the anticipated period of product use is not ascertainable and readily available, the <u>Director</u> may set a minimum period that will be used to calculate average yearly cost of ownership and that will also be part of the product specifications.

(4) The term <u>"Director"</u> means the <u>Director</u> of the (<u>Department of General Services or other similar agency</u>).

(5) The term "public agency" includes an office, department, bureau, board, commission, or agency of the (<u>state or municipality</u>).

Section 5. (a) All bids submitted for contracts to supply energy-consuming commodities to any public agency shall be evaluated according to the life cycle cost of the commodities. Contract award shall be made to the responsible bidder who submits a responsive bid with the lowest average yearly cost of ownership.

(b) The following information shall be determined by the <u>Director</u> for each type of energy-consuming commodity to be acquired by any public agency, and this information shall be set out clearly in invitations to bid for contracts to supply energy-consuming commodities:

(1) The types of costs that will be included in the life cycle cost evaluation;

(2) The method for determining anticipated period of product use, or the minimum period to be applied equally to all bids;

(3) The number of hours of annual operation upon which the calculation of life cycle cost shall be based;

(4) The cost of energy upon which the calculation of life cycle cost shall be based;

(5) The discount rate by which future costs and savings shall be reduced to present value; and

(6) The mathematical formula for determining life cycle cost.

Section 6. [Severability.]

AUTHOR'S SUGGESTED LEGISLATIVE APPROACH #5

IMPORTANT NOTICE: This draft legislative approach constitutes no more than suggestions with respect to the problems posed. It should, therefore, be introduced only after careful consideration of local conditions. Existing constitutional and statutory requirements should be thoroughly examined. Revisions in the statutory language, section headings, numberings, and other modifications may be necessary in order to conform to local law and practices.

Section 1. This act shall be called the "Procurement Specifications Review Act of 197_ ."

Section 2. [Preamble.]

Section 3. For the purposes of this act:

(1) The term "performance-oriented" refers to a specification that is for the procurement of an energy-consuming commodity and that sets out commodity requirements primarily in terms of the way in which the commodity is required to perform.

(2) The term "design-oriented" refers to a specification that is for the procurement of an energy-consuming commodity and that sets out requirements for the construction of the commodity primarily in terms of the materials and design required for the commodity.

(3) The term "energy-consuming commodity" means a product that consumes energy to perform its primary functions.

(4) The term "Director" means the Director of the (Department of General Services or other similar agency.)

(5) The term "public agency" includes an office, department, bureau, board, commission, or agency of the (state or municipality).

Section 4. Within ____ days of the date of enactment of this act, the Director shall develop performance-oriented specifications to govern the purchase by any public agency of energy-consuming commodities. Bid solicitation materials shall state that this use of performance specifications is intended to provide bidders with maximum latitude for innovation in energy-efficient design. Bid solicitation materials shall set out the manner of testing and inspection to be employed by the government in order to ascertain whether the performance-oriented specifications have been complied with.

Section 5. [Optional] There is hereby appropriated the sum of $ _____ for the (establishment/enlargement) of a testing facility within the <u>Department of General Services</u> for the testing and inspection of energy-consuming commodities designed to conform to performance-oriented specifications.

Section 6. [Severability.]

AUTHOR'S SUGGESTED LEGISLATIVE APPROACH #6

IMPORTANT NOTICE: This draft legislative approach constitutes no more than suggestions with respect to the problems posed. It should, therefore, be introduced only after careful consideration of local conditions. Existing constitutional and statutory requirements should be thoroughly examined. Revisions in the statutory language, section headings, numberings, and other modifications may be necessary in order to conform to local law and practices.

Section 1. The <u>Director</u> shall, in the preparation of specifications for government procurement of supplies, ensure that these supplies are recyclable to the maximum extent practicable.

✳ *Chapter 2*

Purchasing Strategies for Reducing Indirect Energy Use

Governments that hope to develop a comprehensive approach to energy conservation must seek to reduce their indirect energy consumption as well as their direct consumption. Direct energy consumption, e.g., energy that flows through an electric or gas meter or into a car's fuel tank, is only part of the energy used by state and local governments. Every item that a government buys represents significant energy investments.

The energy invested in a given item includes energy consumed (1) in extracting or harvesting component raw materials; (2) in refining those materials; (3) in fabricating the finished item; (4) in transporting the components and the finished product to the ultimate consumer; and (5) in recycling or disposing of the item after use [1]. In addition, certain community services, such as trash collection, are frequently provided via government contracts with private operators. The energy used by these operators in the performance of contract services can also be characterized as indirect government energy use.

This chapter presents four discrete procurement strategies by which governments can reduce their indirect energy use:

1. purchasing items made from recycled materials,
2. requiring the transport of government purchases by the most energy-efficient means,
3. procuring services that are performed in an energy-efficient manner, and
4. requiring use of returnable beverage containers at government facilities.

This list of strategies is by no means comprehensive. Given the pervasiveness of energy in the production of goods, the construction of buildings, and the provision of services, the possible strategies are manifold. Other possibilities, not discussed, include purchasing goods that are more durable, that are made from less energy-intensive materials, and that are packaged in reusable containers. The three strategies here are illustrative only, and are also thought to be relatively easy to implement. Strategies explored elsewhere in the book also have relevance to this topic [2].

PURCHASING ITEMS MADE FROM
RECYCLED MATERIALS

The use of recycled, as opposed to virgin, materials in fabricating products is a virtually foolproof way to save energy. The potential savings approximate the difference between the energy required to extract, harvest, and refine component raw materials, and the energy invested to recover secondary materials. One recent study for the Ford Foundation's Energy Policy Project concluded that recovery of ferrous metals, aluminum, and copper from urban scrap would result in energy savings of 86 percent, 96 percent, and 91 percent, respectively [3].

The substantial energy conservation potential of recycling, and its potential for conserving natural resources and reducing solid waste, make the acquisition of products fabricated from recycled materials a very attractive purchasing strategy. To implement the strategy, a government could develop product specifications requiring certain products to be made from recycled materials to the maximum extent practicable [4].

Four questions face governments considering such a strategy:

1. Do existing purchasing specifications preclude the use of recycled materials?
2. Are the costs involved sufficiently low to make this strategy practical?
3. Does limiting government purchases of a particular product to items containing recycled materials so limit the number of bidders as to interfere with competition?
4. What is the maximum practicable level of recycled material content for each type of product?

Although each government must answer these questions for itself, the following discussion will highlight some of the relevant issues.

Existing Specifications

New U.S. Environmental Protection Agency (EPA) guidelines for procurement by the federal government state that purchasing specifications "constitute a barrier to increased resource recovery because recycled materials are often excluded" [5]. Although made in reference to federal specifications, this statement applied to many state and local specifications that were modeled on federal practices.

Specifications that prohibit the use of recycled materials reflect our society's preference for what is new. The improbable justification for incorporating this preference into purchasing specifications is to protect governments from receiving substandard products. In many cases, these specifications are unnecessarily stringent. Thus, in the context of present energy and other resource shortages, governments should reexamine existing specifications. Specifications should be perused for any express prohibitions on the use of recycled materials, and also for performance requirements that are so stringent as to limit unnecessarily the recycled material content [6]. Specifications should also be amended to require the inclusion of the maximum practicable amount of recycled material.

Costs

The administrative costs of purchasing recycled goods include not only the costs of reexamining and revising specifications, but also the costs of determining the maximum practicable level of recycled material content. These costs will arise early in the implementation process, but once they are incurred, few other administrative costs attributable to this strategy will remain [7]. The magnitude of these costs will, of course, vary from government to government. If a government discovers these costs are too great to incur all at once, it may implement the strategy in stages. For example, a given number of specifications could be revised each year, starting with the ones for the most energy-intensive products, until all were revised. Small jurisdictions may have to limit their application of this strategy to the most energy-intensive products to reduce costs to a level they can cope with.

Recycled materials will often cost *more* than virgin materials, which means that a government implementing this strategy will have to pay more to acquire them. Paying these temporarily higher prices is necessary, however, to reverse the prevailing practice of throwing away materials that could have been recycled. This practice of discarding arose as a result of the abundance of energy and other raw materials in the recent past. Government policies have reflected, and often reinforced, this waste [8].

Even though natural resources are now dwindling, market forces and institutional supports still reinforce the use of virgin materials. As more and more natural resources (including energy) are depleted, this situation will gradually reverse itself. In the meantime, however, vast quantities of irretrievable resources may be lost. A government's role in this problem is analogous to fighting a fire. The fire will eventually cease to burn whether or not it is fought, but by fighting it, much can be saved from the flames. In like manner, this government purchasing strategy can help preserve natural resources until the market places greater value on recyclable materials. Government efforts may determine whether there is anything left to recycle when this time arrives.

In some situations products containing recycled materials will already cost less than those containing primarily virgin materials. When overly stringent requirements are discovered and relaxed, it may be discovered that the energy cost savings of using recycled materials are so great that they more than cover the cost of altering production practices. Manufacturers would recover these costs, but they could pass them through to suppliers, and hence to the government, in the form of lower purchase prices.

Interference With Competition

Competition is of utmost importance to the purchasing process, and specifications often critically affect competition [9]. The impact on competition of limiting purchases to products containing recycled materials must be considered on a product-by-product basis. Each purchasing agency must ask whether potential suppliers of a particular product will be deterred from submitting bids if they are required to fabricate the product partly from recycled materials. The answer will depend on several variables. Major factors include the availability of recycled materials to potential suppliers, the ease (or difficulty) with which the use of these materials can be incorporated into existing production routines, and the volume of government purchasing of particular products. This last factor, volume, may determine whether small jurisdictions adopt this strategy, since producers cannot be expected to alter their production processes significantly without an expectation of large sales.

Notifying manufacturers prior to bid solicitation of government intention to require use of recycled materials in purchases, and even requesting their comments, may serve to maintain a high bid response. Also, it may be desirable for several jurisdictions to enter into a cooperative purchase arrangement to generate strong bidder response to revised specifications. The Council of State Governments points out that one reason for such arrangements is "to try to create a demand large enough to encourage the manufacture of new or modified products which are not otherwise commercially available" [10]. Thus, cooperative purchasing is a valuable collateral strategy that could greatly facilitate the implementation by small governments of any purchasing strategy that requires market leverage. (See Chapter 5 for further discussion of cooperative purchasing.)

Maximum Practicable Level of Recycled Material Content

After a government has decided that suppliers will respond to revised specifications, it must determine the maximum practicable level of recycled material in a given product [11]. The EPA guidelines suggest the following considerations:

1. Product performance characteristics
2. Recycled material supplies
3. Material and product costs [12]

These three factors, particularly the second, are often sufficient to determine the maximum level of recycled material content. But a fourth factor of particular importance to small governments is the likely number of bidders. If there are only one or two bidders, they may take advantage of their uniqueness at government expense [13].

Because recycling offers great potential for conserving energy and other natural resources and for reducing solid waste, governments that implement this strategy will make a significant contribution toward solving many societal problems. Not only will they achieve tangible energy savings, but they will also stimulate the demand for resource recovery and will set an example for the private sector.

REQUIRING THE TRANSPORT OF GOVERNMENT PURCHASES BY THE MOST ENERGY-EFFICIENT MEANS

A major buyer of goods can control the mode by which the goods are shipped to him [14]. While adhesion contracts may force private consumers to accept the seller's terms, governments usually have enough buying power to allow them to negotiate terms.

The power to choose the transportation mode for one's purchases can be a potent force for energy conservation. Of a total national energy use of 67,400 trillion BTU in 1970, nearly 10 percent was used to transport freight [15]. While government-purchased freight represents a small part of the total, the aggregate of state and local government freight shipment is significant, and should not be overlooked.

Table 2–1 breaks down the energy efficiency of various modes of freight transport [16].

Obviously, not all these transport modes are available to every purchasing agency in every instance, nor are they all appropriate for every type of good. It has been noted, for instance, that typewriters travel poorly by pipeline. Furthermore, some geographic areas are not served by waterways, and others lack good rail service. When purchasing agencies do have a choice, the table points out how they may save a lot of energy.

A recent study reports that air freight is the fastest growing mode [17]. Given the greater cost of this mode, this increased use must be motivated by a need to receive freight very quickly. In the context of government operations and procurement, this need for fast delivery suggests poor planning. To the maximum extent possible, needs should be forecast. Planning facilitates the use of slower, more efficient transport, as well as volume and cooperative purchasing [18].

Problems of Implementation

The implementation of this strategy should save money. Shifts away from air transport are particularly thrifty. As to modal shifts among railroads, waterways, and trucks, prices are in roughly the same range, although trucks consume far more energy than the other two.

A major problem may be political resistance from representatives of the less efficient modes. They may challenge the purchasing regulations that implement this strategy and claim that the contract provisions hinder competition. Governments must thus draw up regulations that protect or enhance competition while still carrying out

Table 2–1. Energy Efficiency of Freight Transport Modes

Mode	BTU per Ton Mile
Pipelines	450
Railroads	670
Waterways	680
Trucks	2,800
Airplanes*	42,000

*The airplane figure is somewhat overstated, as much freight is shipped on planes that would be carrying passengers, were they available.

this strategy. Another possibility would be to draw up regulations with a stated aim of stimulating the development of an energy-efficient mode of transport. There may be analogies between such a strategy and purchasing techniques designed to aid small businesses [19].

A significant legal problem may arise with respect to liability for the goods in the event of their loss or damage during shipment. If the purchasing government takes charge of shipping the goods, and the delivery term of the contract is F.O.B. the place of shipment, then, under Section 2−319 of the Uniform Commercial Code, the risk of loss or damage during shipment is borne by the government. (When the delivery term is F.O.B. the place of *delivery*, the risk of loss or damage during shipment is borne by the seller.) Since the liability of common carriers (freight transport companies) is generally rather low, a government's bearing the risk of loss or damage could be quite a perilous venture, particularly for a large and valuable shipment. Accordingly, if designating the mode of transport for the goods it purchases requires that a government bear this risk, the purchasing agent should be well aware of it, and will probably wish to obtain insurance on the goods. Of course, this problem can be avoided completely by obtaining a contract that provides for both energy-efficient transport of the goods and the F.O.B. place of destination delivery term.

REQUIRING THAT SERVICES DONE UNDER GOVERNMENT CONTRACT BE PERFORMED IN AN ENERGY-EFFICIENT MANNER

This strategy is quite similar to the one on using contract leverage to influence energy conservation practices by government contractors. There is a slight distinction, as this strategy is intended to apply to services that are performed either within the government or in the government's name, while the other applies to contractors who supply goods to, or construct buildings for, the government. The rationale for making a distinction is that in the former case the contractor's actions are a more integral part of the government's actions; they are often perceived by the public as the government's actions. Thus, government control of the efficiency of these actions (via the contract) is more acceptable, both to the contractor and to the government. This strategy is, accordingly, expected to be easier and cheaper to implement.

There are many services that may fall under the scope of this subchapter: trash collection, public transit, maintenance services, and food services to government buildings, etc. In all cases, the service must be provided by a private contractor, rather than by government employees, for the strategy to apply. For purposes of illustration, this subchapter will discuss only trash collection.

Trash Collection, Suggested Implementation

A recent study for the U.S. Environmental Protection Agency sets out ways to improve the energy efficiency of solid waste collection [20]. These measures include:

1. Less frequent collection,
2. Collection of all wastes simultaneously (except where separate pickup would facilitate solid waste recovery),
3. Improved routing,
4. Use of more efficient and properly sized collection equipment, and
5. Improving storage practices [21].

Energy conservation methods such as these could be implemented in two ways: (1) by writing them into the specifications that are part of the invitation for bids (all bidders would, therefore, have to agree to them), or (2) by requiring bids to be accompanied by "energy usage plans" which would detail the anticipated use of energy in performance of the contract, and the means the contractor would use to limit his energy use. Under the second method, the award of contracts could be based in part on the relative merits of the "energy use plans."

Both methods may foment legal challenges. In the first case, if requirements were so stringent as to eliminate potential bidders, there might be charges of restricting competition. If the second approach is used, there might be charges that contract award was made upon an improper evaluation of energy use plans, since the evaluation of these plans would be fairly subjective.

A safer method, which would nonetheless inform purchasing officials on contractors' capabilities, would be to require the submission of technical proposals prior to the invitation for bids and then base the invitation on the technical proposals. This two step process is acceptable in most jurisdictions [22].

REQUIRING USE OF RETURNABLE BEVERAGE CONTAINERS IN GOVERNMENT INSTALLATIONS

Using recyclable or refillable containers, and facilitating their return to the manufacturer, saves energy. Jettisoning a single beverage bottle wastes as much energy as a 100 watt bulb uses in four hours [23]. Returnable containers will soon be required for beverages sold at federal facilities as a result of the Environmental Protection Agency's promulgation of "Beverage Container Guidelines" [24]. EPA developed these guidelines, which apply to all federal agencies and facilities, to fulfill its responsibilities under the Solid Waste Disposal Act of 1965 [25] as amended by the Resource Recovery Act of 1970 [26].

The objectives of the guidelines are to reduce solid waste and litter and to conserve energy and materials through the use of a return system for beverage containers [27]. Congress views the EPA guidelines as a demonstration effort that will test the impact and implication of a nationwide container deposit law [28]. Consequently, the Senate rejected an amendment to the Solid Waste Utilization Act of 1976 that would have authorized a nationwide container deposit law. Instead, they authorized a study of all aspects of national beverage container legislation. While the EPA regulations affect a small and widely dispersed market [29], the policy deserves discussion here since it may influence state and local government practices.

The requirements section of the guidelines delineates the minimum actions for federal agencies [30]. Section 211 of the Solid Waste Disposal Act and Executive Order 11752 obligate federal agencies to fulfill the requirements spelled out in this section. The EPA recommends that state and local governments, along with private agencies, adopt the guidelines requirements also [31]. Basically, the guidelines require that: (1) federal facilities use only returnable beverage containers, (2) such containers be clearly marked as returnable and labels tell the purchaser where to return the containers, and (3) federal agencies report implementation plans or nonimplementation decisions to EPA regularly.

Reuse or recycling of containers is encouraged via an incentive to return the containers. A 5 cent deposit is levied on each container. The deposit is paid upon purchase by the consumer and refunded by the dealer when the empty container is returned [32]. The EPA also recommends that the federal facility apply higher or lower deposit levels in localities where the established return system calls for higher

or lower deposits [33]. The EPA expects that the deposit requirement will induce consumers to return empty containers.

The EPA guidelines permit some flexibility as far as implementation is concerned, and then allow nonimplementation where compliance with the requirements is impractical because of geographic or
local logistic problems [34]. However, federal agencies that decide
not to use returnable containers must demonstrate to the EPA that
the costs of following the requirements would be excessive and irre-
coverable, and that an alternative cost-effective method of meeting
the guidelines objectives is unavailable [35]. Whether this flexibility
in the guidelines will create an unnecessarily large escape hatch for
federal agencies remains to be seen. The effectiveness of the regulations as a means of reducing federal waste and conserving energy
will depend on how rigorously these provisions are reviewed and en-
forced.

The effect of the EPA regulations on state and local governments
depends partially upon the weight given to EPA's "recommendation" that other levels of government adopt similar requirements
[36]. Perhaps more powerful than this recommendation is the state's
existing legislative or administrative authority to promulgate the necessary regulations [37]. Where such authority does not exist, state
and local government agencies must depend upon the initiative of the
state legislature and upon public lobbying efforts. As evidence of the
feasibility of state legislative action in this area, two states, Oregon
and Vermont, have already enacted mandatory deposits for beverage
containers [38]. Public lobbying efforts have also proved successful
in Massachusetts, Maine, Michigan, and Colorado, where environmental groups generated sufficient support to get the issue submitted by
referendum to the public during the November 1976 election [39].

There are advantages to state legislative enactment of a returnable
beverage container policy as opposed to an administrative promulgation of regulations. First, the state legislature has the authority
to control both the public and private sectors. For example, the
EPA's guidelines are mandatory only as applied to federal facilities
because EPA's authority to promulgate regulations is so limited
under the Resource Recovery Act of 1970. In contrast, both Oregon
and Vermont's deposit requirements are mandatory statewide, in
both the governmental (including federal) and private sectors. Second, state legislatures, since they can enact laws as well as issue more
limited regulations, can make the provisions more stringent and
comprehensive than can administrative agencies. For example, the
Oregon statute provides for "certification" of bottles designed so
that more than one manufacturer can reuse them [40]. These certified bottles have lower deposit rates—an advantage to both the pur-

chaser and the seller [41]. In Vermont, the legislature has prohibited the sale of any metal beverage container that requires detaching a part to open the container or of any glass container that cannot be refilled [42]. However, Vermont's prohibition of "pop tops" does not necessarily signal a trend; several comments challenging EPA's proposed guidelines were based on the fear that these regulations would bar certain types of containers [43]. In the final regulations, the EPA carefully and clearly pointed out that the regulations did not require the use of refillable bottles, but only required the use of returnable bottles.

But what can state or local governments do to implement a container return policy if the state legislature fails to enact comprehensive legislation? In many states, legislative authority may already exist to issue "necessary" regulations in the area of solid waste management. Many state resource recovery acts or solid waste acts authorize broad administrative action similar to that taken by the EPA. While such legislation may not allow regulation of both the public and private sector, it may authorize the development of EPA type regulations applicable to state or local governmental agencies. Since most governmental agencies have authority to control their internal operations and administrative procedures, guidelines could perhaps be issued strictly on the basis of that authority.

It is hoped that state and local governments will implement these strategies, where appropriate, and that this will serve as a first step toward reducing indirect governmental energy use. Under certain circumstances, strategies of the type discussed in this chapter will save government money, as well as energy. In other situations, financial incentives may derive from the State Energy Conservation Plans of the Energy Policy and Conservation Act [44]. Energy savings attributable to strategies that reduce indirect energy use may be used in calculating the 5 percent reduction in state energy consumption which the act sets as a goal for the conservation plans [45].

Persons using the approaches of this chapter to develop and implement conservation strategies are urged to exercise their imagination and creativity. Indirect energy use is extremely pervasive, and the four strategies presented here represent only the tip of the iceberg.

NOTES TO CHAPTER 2

1. *See*, Clark W. Bullard III and Robert A. Herendeen, Energy Impact of Consumption Decisions, *Proceedings of the IEEE* [Institute of Electrical and Electronics Engineers, Inc.] 63, no. 3 (March 1975): 484−93.

2. These include strategies presented in Chapters 3 and 7.

3. William E. Franklin, David Bendersky, William R. Park, and Robert G. Hunt, "Potential Energy Conservation from Recycling Metals in Urban Solid Waste," in *The Energy Conservation Papers*, ed. Robert H. Williams (Cambridge, Massachusetts: Ballinger, 1975), p. 172.

4. *See* appendix to this chapter for (1) an excerpt from Pub. L. No. 94—580 (October 21, 1976),which requires federal procurement,with limited exceptions, of items composed of the highest percentage of recovered materials practicable; (2) Me. Rev. Stat. Ann. tit. 5, §1812, *as amended* (1976), which requires, under certain conditions, the purchase of supplies and materials "which are composed in whole or in part of recycled materials"; and (3) Maryland House of Delegates, Bill No. 1097 (introduced January 29, 1976), which requires a scaled introduction of recycled paper into state purchasing.

5. 41 Fed. Reg. 2356 (January 15, 1976). *See also, Guidelines Issued Pursuant to the Solid Waste Disposal Act*, 6 ELR 10044 (1976); and Environmental Law Institute, *Federal Environmental Law*, ed. Erica L. Dolgin and Thomas P. Guilbert (St. Paul, Minnesota: West, 1974), p. 1304.

6. 41 Fed. Reg. 2357—58 (January 15, 1976).

7. Costs that will recur include those of periodically reviewing the specifications once they are initially revised. This cost is not viewed as particularly burdensome, since some forms of specification review should be periodically undertaken in any case.

8. *See, Federal Tax Policy Has Only Modest Impact on Recycling, Environmental Law Institute Study Concludes*, 6 ELR 10041 (1976). This article sets out a variety of government policies that appear to favor the use of virgin materials over recycled materials and that thereby increase the cost of recycled materials relative to virgin materials. Those policies include freight rates, source labeling requirements, former federal procurement policies, federal mineral discovery policy, and municipal waste removal subsidies.

9. For discussions of impediments to competition and of the intricacies of specification writing, *see* The Council of State Governments, *State and Local Government Purchasing* (Lexington, Kentucky, 1975), pp. 7.1—7.10 and 11.1—11.2.

10. *Ibid*, p. 14.1. Note also that the National Institute of Governmental Purchasing, 1001 Connecticut Avenue, N.W., Washington, D.C. 20036, is actively engaged in assisting local governments to participate in cooperative purchasing arrangements.

11. See appendix to this chapter for a U.S. Federal Supply Service (FSS) Fact Sheet that sets out percentages of reclaimed materials required in FSS procurements.

12. 41 Fed. Reg. 2356 (January 15, 1976).

13. Again, this lack of vigorous competition is due to the relatively low volume of purchasing for a single state or local government as compared with the federal government. (Exceptions should be noted in the case of large states, such as New York and California, and large cities.)

14. Section 2—319 of the Uniform Commercial Code governs the allocation of risk and expense of shipment between buyer and seller.

15. Eric Hirst, Transportation Energy Conservation: Opportunities and Policy Issues, *Transportation Journal*, 13, no. 3 (Spring 1974): 43.

16. Eric Hirst and Robert Herendeen, "Total Energy Demand for Automobiles," Society of Automotive Engineers Publication 730065 (1973). Other tables setting out the energy intensiveness of the several freight modes are found in the appendix to this chapter.

17. Denis Hayes, "Energy: The Case for Conservation," *Worldwatch Paper 4* (Washington, D.C., January 1976), p. 31.

18. *See* Chapters 1 and 5 of this book.

19. *See e.g.*, Federal Procurement Regulations, 41 C.F.R. Subpart 1–1.7 (Small Business Concerns).

20. U.S. Environmental Protection Agency, *Energy Conservation Through Improved Solid Waste Management*, prepared by Robert A. Lowe, with appendices by Michael Loube and Frank A. Smith (Washington, D.C., 1974), pp. 33–35. As discussed in Chapter 8, resource recovery from solid waste has a tremendous energy conservation potential. Here, however, our topic is limited to the *collection* of solid waste.

21. *Ibid.*, pp. 34–35.

22. *See* Federal Electric Corporation v. Fasi, 527 P.2d 1284 (1974).

23. State of Oregon, Department of Energy, "Family Energy Watch Calendar" (1976).

24. 40 C.F.R. Part 244, 41 Fed. Reg. 41202 (September 21, 1976).

25. Pub. L. No. 89–272, 42 U.S.C. §§ 3251 et seq., ELR 41901.

26. Pub. L. No. 91–512, 42 U.S.C. §§ 3251–54f, 3256–59 (1970), ELR 41902.

27. 40 C.F.R. Part 244.100(c) (1) and (d) (1).

28. 122 Cong. Rec. S11058–86 (daily ed., June 30, 1976).

29. Federal government facilities comprise only 2–4 percent of the national beverage market. 41 Fed. Reg. 41202 (September 21, 1976).

30. 40 C.F.R. Part 244.200.

31. 40 C.F.R. Part 244.100(b).

32. 40 C.F.R. Part 244.100(c) (2).

33. 40 C.F.R. Part 244.100(c) (3).

34. 40 C.F.R. Part 244.100(d).

35. 40 C.F.R. Part 244.100(d) (3).

36. 40 C.F.R. Part 244.100(b).

37. Conversation with Mr. Webster, Vermont Environmental Conservation Agency, September 29, 1976. This was cited as the reason why the state government had not issued regulations prior to the legislative enactment.

38. Vt. Stat. Ann. tit. 10, § 1521 et seq. (Supp. 1975); Ore. Rev. Stat. § 459. 810 et seq. (Supp. 1974).

39. Conversation with Mr. William Ades, Office of Solid Waste Management, U.S. Environmental Protection Agency, September 29, 1976.

40. Ore. Rev. Stat. § 459.860 (Supp. 1974).

41. Ore. Rev. Stat. § 459.820(2) (Supp. 1974).

42. Vt. Stat. Ann. tit. 10, § 1525 (Supp. 1975).

43. *See*, 5 ELR 10197–10199 (1975); and 41 Fed. Reg. 41202 (September 21, 1976).

44. Energy Policy and Conservation Act §§ 361–366, 42 U.S.C. §§ 6321–6326 (1975).

45. 42 U.S.C. § 6322.

Appendices to Chapter 2

EXCERPT FROM THE RESOURCE CONSERVATION
AND RECOVERY ACT OF 1976, PUB. L. NO. 94–580
(OCT. 21, 1976).

This provision requires federal procurement, with limited exceptions, of items composed of the highest percentage of recovered materials practicable:

Federal Procurement

Sec. 6002. (a) *Application of Section*—Except as provided in subsection (b), a procuring agency shall comply with the requirements set forth in this section and any regulations issued under this section, with respect to any purchase or acquisition of a procurement item where the purchase price of the item exceeds $10,000 or where the quantity of such items or of functionally equivalent items purchased or acquired in the course of the preceding fiscal year was $10,000 or more.

(b) *Procurement Subject to Other Law.*—Any procurement, by any procuring agency, which is subject to regulations of the Administrator under section 6004 (as promulgated before the date of enactment of this section under comparable provisions of prior law) shall not be subject to the requirements of this section to the extent that such requirements are inconsistent with such regulations.

(c) *Requirements.*—(1) (A) After two years after the date of enactment of this section, each procuring agency shall procure items composed of the highest percentage of recovered materials practicable consistent with maintaining a satisfactory level of competition. The decision not to procure such items shall be based on a determination that such procurement items—

(i) are not reasonably available within a reasonable period of time;

(ii) fail to meet the performance standards set forth in the applicable specifications or fail to meet the reasonable performance standards of the procuring agencies; or

(iii) are only available at an unreasonable price. Any determination under clause (ii) shall be made on the basis of the guidelines of the Bureau of Standards in any case in which such material is covered by such guidelines.

(B) Agencies that generate heat, mechanical, or electrical energy from fossil fuel in systems that have the technical capability of using recovered material and recovered-material-derived fuel as a primary or supplementary fuel shall use such capability to the maximum extent practicable.

(C) Contracting officers shall require that vendors certify the percentage of the total material utilized for the performance of the contract which is recovered materials.

(d) *Specifications.*—(1) All Federal agencies that have the responsibility for drafting or reviewing specifications for procurement item procured by Federal agencies shall, in reviewing those specifications, ascertain whether such specifications violate the prohibitions contained in subparagraphs (A) through (C) of paragraph (2). Such review shall be undertaken not later than eighteen months after the date of enactment of this section.

(2) In drafting or revising such specifications, after the date of enactment of this section—

(A) any exclusion of recovered materials shall be eliminated;

(B) such specification shall not require the item to be manufactured from virgin materials; and

(C) such specifications shall require reclaimed materials to the maximum extent possible without jeopardizing the intended end use of the item.

(e) *Guidelines.*—The Administrator, after consultation with the Administrator of General Services, the Secretary of Commerce (acting through the Bureau of Standards), and the Public Printer, shall prepare, and from time to time revise, guidelines for the use of procuring agencies in complying with the requirements of this section. Such guidelines shall set forth recommended practices with respect to the procurement of recovered materials and items containing such materials and shall provide information as to the availability, sources of supply, and potential uses of such materials and items.

(f) *Procurement of Services.*—A procuring agency shall, to the maximum extent practicable, manage or arrange for the procurement of solid waste management services in a manner which maximizes energy and resource recovery.

(g) *Executive Office.*—The Office of Procurement Policy in the Executive Office of the President, in cooperation with the Administrator, shall implement the policy expressed in this section. It shall be the responsibility of the Office of Procurement Policy to coordinate this policy with other policies for Federal procurement, in such a way as to maximize the use of recovered resources, and to annually report to the Congress on actions taken by Federal agencies and the progress made in the implementation of such policy.

AMENDMENT TO ME. REV. STAT. ANN. TIT 5, §1812 (APRIL 9, 1976):

Sec. I. 5 MRSA §1812, 1st ¶ is amended by adding at the end the following new sentences:

Whenever supplies and materials are available for purchase which are composed in whole or in part of recycled materials and are shown by the seller, supplier or manufacturer to be equal in quality and are competitively priced, the State Purchasing Agent shall purchase such recycled supplies and materials. For the purposes of this section, recycled materials means materials that are composed in whole or in part of elements that are reused or reclaimed.

A BILL BEFORE THE MARYLAND HOUSE OF DELEGATES THAT WOULD REQUIRE THE PURCHASE OF RECYCLED PAPER:

House of Delegates No. 1097

By: Delegates Kernan and Brown
Introduced and read first time: January 29, 1976
Assigned to: Environmental Matters

A BILL ENTITLED

AN ACT concerning

Recycled Paper

FOR the purpose of providing definitions for the terms "recycled paper" and "secondary waste material"; providing certain estimates for the purchase of recycled paper by this State; and renumbering as appropriate.

BY repealing and reenacting, with amendments, Article 41—Governor-Executive and Administrative Departments, Section 231—IA, Annotated Code of Maryland, (1971 Replacement Volume and 1975 Supplement).

Section 1. BE IT ENACTED BY THE GENERAL ASSEMBLY OF MARYLAND, That Section 231—IA of Article 41—Governor-Executive and Administrative Departments, of the Annotated Code of Maryland (1971 Replacement Volume and 1975 Supplement) be and it is hereby repealed and reenacted, with amendments, to read as follows:

Article 41—Governor-Executive and Administrative Departments

231—IA.

(A) As used in this section "recycled paper" means any paper product with not less than 50 percent of its total weight consisting of secondary waste materials.

(B) "Secondary waste materials" means fragments, products, or finished products of a manufacturing process which has converted a raw material into a commodity of real economic value, and includes post consumer waste.

(C) The total paper purchase by the Secretary of General Services shall include recycled paper so that the total purchase of recycled paper by the state is not less than 5 percent by 1977, 10 percent by 1979, 20 percent by 1981, and 35 percent by 1985.

(D) In purchasing any paper or paper products as materials and supplies for the using authorities pursuant to the authority granted by §231G of this article, the Secretary of General Services, to the fullest extent practicably possible, shall purchase or approve for purchase only such materials and supplies that are manufactured or produced from recycled paper.

Section 2.   AND BE IT FURTHER ENACTED, That this Act shall take effect July 1, 1976.

Explanation: Capitals indicate matter added to existing law. [Brackets] indicate matter deleted from existing law. Numerals at right identify computer lines of text.

FEDERAL SUPPLY SERVICE (FSS) FACT SHEET SETTING OUT PERCENTAGES OF RECLAIMED MATERIALS REQUIRED IN CERTAIN FSS PROCUREMENTS, MARCH 1976*

Fact Sheet: Recycled Materials

In 1973, over 62 million tons of paper and paperboard were used in the United States. Economists project that by 1985, consumption of these products will increase to 100 million tons per year. This rapid increase, together with the expansion of our cities, increasing scarcity of sites for sanitary land fills, and the phaseout of incineration due to air pollution regulations, contribute to the scope and severity of our solid waste problem. In view of the projected increase in waste generation, the amount of reclaimed fibers used must be substantially increased to restrict the volume of materials entering the solid waste stream.

Aware of the problem, the General Services Administration, at the direction of the President, has undertaken a role of national leadership in the promotion of recycling solid waste for the production of new items. Recycled waste material contributes to the economy and can be looked upon as a resource rather than a liability.

About 50 percent of urban waste consists of paper and paperboard products.

---

*Source: Hearings on H.R. 14496 Before the Subcommittee on Transportation and Commerce of the House Committee on Interstate and Foreign Commerce, 94th Cong., 2d sess., at 121–130 (1976). Note that H.R. 14496 was enacted into law on October 21, 1976, as the Resource Conservation and Recovery Act of 1976, Pub. L. No. 94–580 (Oct. 21, 1976).

This fact points out the significance of GSA's paper recycling program. Many common use paper products purchased for Federal agencies by GSA are now required to contain percentages of recycled fibers. The percentages vary from 3 to 100. It is expected that these requirements, which are implemented with the cooperation of industry, will promote recycling programs at all levels of government and in private enterprise.

Because the purpose of our program is to divert as much waste as possible from incinerators and land fills, GSA requires the use of post-consumer waste in its recycled paper products whenever feasible. Post-consumer waste is defined as material that has already been circulated commercially and used and discarded from homes, offices, and factories.

Reclaimed fibers applicable to the paper products GSA procures are defined in Enclosure 1. Enclosure 2 lists the percentages of recycled materials currently required for specific products. These percentages do not represent GSA's final position, but are steps in a progressive approach toward achieving maximum recycling. Enclosure 3 lists, by commodity and Federal Specification number, the names and addresses of suppliers who have offered some percentage of reclaimed fibers when submitting bids.

To date, reclaimed fibers are required in products covered by 86 purchase specifications, including such categories as:

1. Building materials—6 specifications.
2. Office supplies—36 specifications.
3. Packaging—33 specifications.
4. Tissue—7 specifications.
5. Miscellaneous items—4 specifications.

Enclosure One: Definition of Reclaimed Fibers Currently Used in Procurement

The paper stock shall contain not less than _____ percent, by weight, of fibers listed in Part I and/or Part II below that are reclaimed from solid waste or waste collected as a result of a manufacturing process but shall not include those materials generated from and reused within a plant as a part of the papermaking process.*

Where specific percentages of reclaimed fibers from Part I are required, the following will be added to the definition:

A minimum of _____ percent of the total weight of the paper stock shall be of reclaimed fibers from sources listed in Part I. The remainder of the reclaimed fibers shall be from sources listed in Part II.

In all cases, a certificate shall be submitted with each bid, specifying the percentage of reclaimed fibers to be used in the manufacture of the item offered to

---

*The papermaking process is defined as those manufacturing operations up to and including the cutting and trimming of the paper machine reel into smaller rolls or rough sheets.

the Government. The certificate shall indicate the percentage of the total fiber content, by weight, to be supplied from sources listed in Part I, and the percentage of the total fiber content to be supplied from sources listed in Part II.

Part I

A. Paper, paperboard, and fibrous wastes from factories, retail stores, office buildings, homes, etc., after they have passed through their end-usage as a consumer item, including:

1. Used corrugated boxes;	4. Mixed waste paper;
2. Old newspapers;	5. Tabulating cards; and
3. Old magazines;	6. Used cordage.

B. All paper, paperboard, and fibrous wastes that enter and are collected from muncipal solid waste.

Part II

A. Dry paper and paperboard waste generated after completion of the paper-making process* including:

1. Envelope cuttings, bindery trimmings, and other paper and paperboard waste, resulting from printing, cutting, forming, and other converting operations;
2. Bag, box, and carton manufacturing wastes; and
3. Butt rolls, mill wrappers, and rejected unused stock.

B. Finished paper and paperboard from obsolete inventories of paper and paperboard manufacturers, merchants, wholesalers, dealers, printers, converters, or others.

C. Fibrous by-products of harvesting, manufacturing, extractive or woodcutting processes, flax straw, linters, bagasse, slash and other forest residues.

D. Wastes generated by the conversion of goods made from fibrous materials; i.e., waste rope from cordage manufacture, textile mill waste and cuttings.

E. Fibers recovered from waste water which otherwise would enter the waste stream.

Enclosure Two: Percentages of Reclaimed Materials Required in GSA Procurements

Specification Number	Commodity	Minimum Percent Reclaimed Material
	BUILDING MATERIALS	
HH–I–515B	Insulation, thermal, batt & blanket	Type I–100 (90PCW[†]); Type II–80 (40PCW)
HH–I–1030A	Insulation, thermal, mineral fiber (Types I and II)	15

*The papermaking process is defined as those manufacturing operations up to and including the cutting and trimming of the paper machine reel into smaller rolls or rough sheets.

[†]PCW = Post-consumer waste.

Specification Number	*Commodity Building Materials (cont'd.)*	*Minimum Percent Reclaimed Material*
HH–R–595B	Roofing felt, coal tar & asphalt saturated	40 (30PCW)
SS–R–501D	Roofing felt, asphalt-prepared, smooth surfaced	40 (30PCW)
SS–R–630D	Roofing felt, roll, asphalt-prepared, mineral surfaced	40 (30PCW)
LLL–I–535A	Insulation, board & block, thermal	15

OFFICE SUPPLIES

GG–H–1314A	Label, holder, binder	25
GG–H–1488A	Label, holder, desk tray	25
SS–P–00196D	Pencil, paper wrapped	35
UU–B–00320D	Binder, loose leaf	100 (75PCW): liner–40
UU–B–336D	Binder, loose leaf, flexible post	100 (75PCW); liner–40
UU–B–338C	Binder, loose leaf, flexible tape	100 (75PCW); liner–40
UU–B–340A	Binder, loose leaf, flexible prong	25
UU–B–00344	Binder, multiple ring	100 (75PCW)
UU–B–356A	Binder, loose leaf, ring	100 (75PCW); liner–40
UU–B–00366B	Binder, loose leaf, acc. post	100 (75PCW); liner–40
UU–B–368A	Binder, loose leaf, spring back	100 (75PCW); liner–40
UU–B–00400A	Binder, note pad (NIB item)	100 (75PCW)
UU–B–690	Filing box, letter	90 PCW
UU–C–95B	Guide cards	50 (10PCW); except pressboard–35 (10PCW)
UU–C–00250B	Filing case	Inner binder 100PCW; outer case 40PCW
UU–C–255A	Filing case, transfer, fiberboard	35 (28PCW)
UU–D–650A	Doily, paper	50 (40PCW)
UU–E–522E	Mailing envelope	25 white bond; light colors–20; brown-hued colors–30
UU–F–00350B	File backer	17
UU–F–1206A	File folder	Grade A–35 (10PCW); Grades B, C, & D–20
UU–L–0049C	Label, paper, gummed	25 (10PCW)
UU–P–12C	Columnar pad	Sheets 25; backing 100PCW
UU–P–0016B	Pad, desk blotter	Pad 100 (75PCW); facing sheets–20
UU–P–21F	Pad, writing paper (memo pad)	Pad 20; backing 100 (50PCW)
UU–P–63J	Blotting paper	35

Specification Number	Commodity *Office Supplies (cont'd.)*	Minimum Percent Reclaimed Material
UU–P–202D	Graph paper	50 (Grade C only)
UU–P–320B	Loose leaf paper, ruled and blank	30
UU–P–547J	Teletypewriter paper	20 (Grade B only)
UU–P–561H	Paper, tracing	100 (except Type II)
UU–P–663B	Paperboard, drawing	85 (Core only)
UU–P–675B	Pocket planning set and refill	Box 100 (75PCW); filler sheets– 20
UU–P–680A	Portfolio, double pocket, presentation	25
UU–T–110B	Computing machine tape	60 (10PCW)
LLL–B–00650B	Filing box, index card	100 PCW
LLL–C–00110A	Filing case, transfer	75 (25PCW)
LLL–C–001491	Filing case, transfer, knockdown	100 PCW

PACKAGING

UU–B–23C	Bag, paper, foil, laminated, bedside	15
UU–B–25C	Bag, paper insulated	15
UU–B–36J	Bag, paper grocers	25
UU–B–39D	Bag, paper merchandise	15
UU–B–43D	Bag, paper, waste receptacle	15 (5PCW)
UU–B–0050D	Sandwich bag	10
UU–C–225A	Egg carton	100
UU–C–282D	Chipboard	100 (50PCW)
UU–P–134E	Paper, wrapping, wet-waxed	10
UU–P–268G	Kraft paper	25
UU–P–270F	Paper, wrapping, waxed, dry	10
UU–P–272C	Paper, wrapping, freezer	20
UU–P–1382B	Paper, meat interleaving	10
UU–T–81H	Tag, shipping and stock	13
UU–T–665E	Trays and boards, prepacking	30
PPP–B–566E	Box, folding, paperboard	80 (40PCW)
PPP–B–591B	Box, shipping, fiberboard, wood-cleated	3
PPP–B–636H	Box, fiberboard, shipping	35 (10PCW)
PPP–B–640D	Box, fiberboard, triple-wall	25 (5PCW)
PPP–B–650B	Box, fiberboard, record retiring	35 (10PCW)
PPP–B–665D	Box, paperboard, metal-edged	75 (40PCW)—Styles A, B, C, D & E only
PPP–B–1055B	Barrier material, waterproof	30
PPP–B–1364C	Box, corrugated, fiberboard, double-wall	25 (5PCW)

Specification Number	Commodity Packaging (cont'd.)	Minimum Percent Reclaimed Material
PPP–B–001606A	Box, fiberboard, special use	35 (10PCW)
PPP–B–001608	Box, corrugated, fiberboard, weather resistant coated	3
PPP–B–001672	Box, shipping, reusable & cushioning	35 (10PCW)
PPP–C–843C	Cushioning material, cellulosic	70 (30PCW)
PPP–E–540C	Envelope, water resistant	30
PPP–F–320D	Fiberboard, corrugated and solid	35 (10PCW) except triple-wall–25 (5PCW)
PPP–P–291D	Paperboard, wrapping & cushioning	60 (30PCW)
PPP–P–700B	Protector, packing list	35 (10PCW)
PPP–S–30C	Shipping sack	40 (10PCW)
PPP–T–495B	Mailing and filing tube	90 PCW

TISSUE

UU–C–1229	Cover, toilet seat	40
UU–N–1650A	Napkin, table, paper	60 (30PCW)
UU–P–556J	Paper, toilet tissue	50 (20PCW)
UU–T–450D	Tissue, facial	20 (5PCW)
UU–T–591E	Towel, paper	95 (40PCW)
UU–T–595B	Towel, wiping, paper: institutional and industrial	20
UU–T–00598	Towel, paper (plastic-wiping)	20

MISCELLANEOUS

UU–B–838D	Butter chip, paper	25
UU–C–806H	Cups and lids, paper, disposable	15
UU–P–665B	Plate, baking, pie (paper)	15
UU–P–670D	Plate, paper, rectangular and round	15

AUTHOR'S SUGGESTED LEGISLATIVE APPROACH #7*

IMPORTANT NOTICE: This draft legislative approach constitutes no more than suggestions with respect to the problems posed. It should, therefore, be introduced only after careful consideration of local conditions. Existing constitutional and statutory requirements should be thoroughly examined. Revisions in the statutory language, section headings, numberings, and other modifications may be necessary in order to conform to local law and practices.

*Derived almost entirely from U.S. Environmental Protection Agency regulations, 41 Fed. Reg. 2356–58 (January 15, 1976).

Section _____ . The <u>Director</u> shall undertake within _____ days of the effective date of this act an examination of existing specifications used in the procurement of supplies for the government. Existing specifications shall be amended to delete, except where deemed essential to the proper functioning of the product:

(1) express prohibitions on the use of recycled materials;

and

(2) performance requirements that are so stringent as to limit unnecessarily the recycled material content.

In addition, specifications used in the procurement of supplies for the government shall be amended to require the maximum practicable level of recycled material content. This level shall be determined by reference to:
(1) product performance characteristics;
(2) recycled material supplies;
(3) material and product costs; and
(4) the likely effect upon competition for contracts to supply the product of requiring a particular level of recycled material content.

Figure A2–1.   Fuel Consumption of Intercity Freight Movements, by Mode

Mode	Net Ton-Miles Per Gallon	Millions of Gallons 1970
Rail	202	3,827
Truck		
For hire	75	2,933
Private	50	2,640
Water*	220	2,664
Oil Pipeline	500	806
Air	12	325
Total Intercity	160.5	13,195

*Includes domestic deep sea.
Source: Peat, Marwick, Mitchell and Company, *Industrial Energy Studies of Ground Freight Transportation*, vol. I (Washington, D.C.: FEA, 1974), p. 47.

Figure A2–2. Historical Variation in Energy Intensiveness for
Intercity Freight Modes, Plotted Semilogarithmically

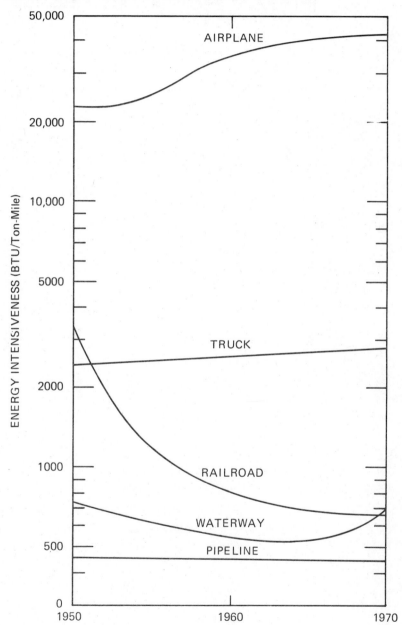

Source: Oak Ridge National Laboratories

 *Chapter 3*

Using Contract
Leverage to Influence
Government Contractors
to Conserve Energy

INTRODUCTION

As previously indicated, strategies in this chapter are intended to be applied to government contractors that supply goods or construct buildings. The goal of the strategies
is to require contractors to adopt certain energy-conserving methods
prescribed by the state or local governments. Adoption of these
methods should be required for the life of the contract and, to reduce
the burden on the contractor to realistic dimensions, for only those
operations relating to performance of the contract.

There are two basic ways of applying leverage: (1) by requiring
suppliers and construction companies to conform their operations to
a certain level of energy conservation to qualify as bidders, and (2)
by inserting provisions in the contract that require that the project
be executed in compliance with certain energy conservation standards.

QUALIFICATION AND DEBARMENT
OF BIDDERS BASED ON THE ENERGY
EFFICIENCY OF THEIR OPERATIONS

According to a recent study by the Council of State Governments, at
least forty states require that suppliers meet certain qualifications
before they are permitted to bid for government contracts [1]. The
study also points out that prequalification of bidders is an increas-

ingly common practice [2]. Potential bidders are generally qualified on the basis of their finances, capability to perform certain types of contracts, and business reputations. Prequalification can reduce the cost of bid solicitation by reducing the size of the bidders list [3]. It can also speed bid evaluation and contract award by eliminating the need to investigate the bidder's qualifications during or after the solicitation process [4]. Further, prequalification can reduce the need for bonding by bidders, which in turn can reduce contract costs to the government [5].

Ensuring that government contractors are reliable is not the only goal to which the prequalification of bidders technique can be applied. To the extent that doing business with a particular contractor represents governmental acceptance of that contractor's practices, governments may wish to ensure that its contractors' practices are socially beneficial, or at least not socially harmful. The bidder prequalification method may be employed to achieve this goal.

Two interesting, and highly similar, applications of this general strategy are found in federal regulations relating to the environment: the Clean Air Amendments of 1970 [6] and the Federal Water Pollution Control Act Amendments of 1972 [7]. Essentially, these acts prohibit federal agencies from contracting with any person who is in violation of the respective act [8]. Since the acts already provide civil and criminal penalties for violation, the procurement sanction is collateral [9]. In addition, the prohibition on contracting is limited to the facility at which violation of the respective act occurred [10]. In other words, a contractor is not wholly barred from government contracting unless its entire operation is in violation of one or both acts.

Application by state and local governments of the bidders qualification strategy to encourage suppliers and contractors to conserve energy is simple enough in theory, but who is to establish the requirements for qualification? Either the state department of general services, or another agency responsible for purchasing and public works, must develop the standards, or they must be "borrowed" from a preexisting enforcement scheme.

If the department of general services is charged with developing the energy conservation standards for bidders, it should seek the assistance of the state energy agency. The energy agency can suggest which general types of standards are likely to achieve the greatest energy saving, and the department of general services can then assess the likely effect of such requirements upon the procurement process. Where state energy agencies have already drawn up voluntary standards for industrial energy conservation, the incorporation of those standards into bidder qualifications would expedite the implementation of this strategy [11].

A state or local government may decide that the energy conservation benefits of such a strategy are insufficient to justify the administrative costs of setting and enforcing standards to implement it. In that case, consideration should be given to the type of use made of the strategy by the federal acts mentioned above, i.e., as a collateral means of enforcing an existing regulatory scheme. One scheme that should soon be implemented in many states, and which could serve as the foundation for a collateral enforcement strategy, is that provided in the Energy Policy and Conservation Act [12]. Subchapter II, Part B, of this act provides for federal funding for state energy conservation plans designed to achieve a reduction in state energy consumption of 5 percent or more from that projected for 1980 [13]. The act requires that these state plans must include five basic elements, of which three could conceivably apply to industrial operations:

(1) mandatory lighting efficiency standards for public buildings (except public buildings owned or leased by the United States government);
(2) programs to promote the availability and use of car pools, vanpools, and public transportation (except that no federal funds provided under this part shall be used for subsidizing fares for public transportation); ...
(4) mandatory thermal efficiency standards and insulation requirements for new and renovated buildings (except buildings owned or leased by the United States); ... [14].

In addition, the federally funded state energy conservation plans *may* include:

(1) restrictions governing the hours and conditions of operation of public buildings (except buildings owned or leased by the United States government);
(2) restrictions on the use of decorative or nonessential lighting;
(3) transportation controls; ... [15].

Once these plans are in effect, debarment from bidding could be used as a collateral sanction against potential contractors that fail to comply with the regulations that implement the state plan.

Qualification and debarment of bidders on the basis of the energy efficiency of their operations is a strategy with pervasive effect. Since bidder qualification is the threshold to government contracting, the strategy could conceivably affect suppliers and other businesses that are never actually awarded a contract. Consequently, to be effective the strategy would have to impose a minimum burden upon would-be bidders. It is suggested that the emphasis in energy conservation regulations be directed toward saving costs to the bidder, as well as toward saving energy, and that enforcement be accompanied with a public relations effort to highlight this dual benefit. Otherwise, if requirements pose an excessive financial burden, suppliers and builders may merely solicit business elsewhere—to the detriment of competition for government contracts.

To implement this strategy, contracts made by government purchasing and public works agencies should require certification that the contractor is in compliance with the regulations at the time of bidding. Contracts should also require continuing compliance with regulations [16]. These contractual measures will serve to clarify the contractor's obligation and to facilitate a remedy against the contractor in the event of noncompliance.

CONTRACT CLAUSES REQUIRING ENERGY CONSERVATION PRACTICES

Many government contracts at federal, state, and local levels include provisions intended to further social goals that go well beyond the subject of the contract. The contract may be for the supply of toilet bowls, but the social goal provisions may require that the bowls be manufactured from materials mined or produced in the United States; [17] that the contractor guarantee equal employment opportunity; [18] that the contractor agree to pay at least the prevailing minimum wage, to require only a forty-hour work week, and to provide a safe and sanitary workplace; [19] or that the contractor give employment preference to veterans [20]. None of these provisions will have any effect upon the product supplied, except perhaps to increase its price, and yet all are deemed justified on social policy grounds [21].

Similarly, it is possible for governments to require contractors to adopt a wide variety of energy conservation measures. Since most contractors utilize buildings and have motor vehicles, as do governments, many requirements can be derived from strategies for state and local governments discussed elsewhere in this book. In addition, *Energy Efficiency in Industry: A Guide to Legal Barriers and Opportunities*, by Norman L. Dean, provides a detailed discussion of industrial energy conservation methods that could be adapted to contract provisions [22].

Federal Preemption

Where there is federal regulation of a given area, states may be precluded from exercising regulatory authority in that area. State action is said to be "preempted" by federal action. This rule is founded upon the Supremacy Clause (art. VI, cl. 2) and the Commerce Clause (art. 1, §8, cl. 3) of the United States Constitution. A recent example of its application is found in the *Northern States Power Co.* v. *Minnesota* case, in which states were precluded from regulating construction and operation of nuclear power plants [23].

The Energy Policy and Conservation Act contains provisions that expressly preempt state regulation of automotive fuel economy and certain aspects of energy efficiency of certain consumer products other than automobiles [24]. (Relevant sections of this act are reproduced in the appendix.) While the act provides specific exemptions from the preemption for procurement by states and their political subdivisions, it is unlikely that these exemptions would extend to contractor compliance strategies in the preempted areas [25]. A state or locality might be precluded from using a contract clause that would require contractors to use fuel-efficient automobiles or certain other energy-efficient products.

In fact, it seems certain that state or local use of contract clauses requiring contractors to use fuel-efficient automobiles would be precluded. (Note that under the act, "automobiles" are not always automobiles) [26]. The applicable section of the act states broadly: "*[N]o* State or political subdivision of a State shall have authority to adopt or enforce *any law or regulation* relating to fuel economy standards or average fuel economy standards . . ." [27]. The procurement exemption is narrowly limited to ". . . automobiles procured for its [a state's or its political subdivision's] own use" [28]. Thus, it would seem necessary to limit any contract clauses relating to motor vehicle fuel efficiency to those vehicles not covered by the act.

As to consumer products other than automobiles, the Energy Policy and Conservation Act is more complex and less clear [29]. The following brief discussion of this part of the act is provided only to orient the reader to the issues. The discussion is general and oversimplified, and the legal resolution of a case arising under this part of the act might well be based upon details that are not here mentioned. Readers interested in implementing energy conservation strategies that might conceivably conflict with this act are thus *strongly urged* to scrutinize the law itself.

The Act's "Energy Conservation Program for Consumer Products Other Than Automobiles" has several components:

1. A section prescribing what type of consumer products are covered by the program [30].
2. A section prescribing the development of test procedures to determine the energy operating costs of covered products and prohib-

iting manufacturers and other sellers of covered products from making representations as to the energy consumption of these products unless the products have been properly tested [31].

3. A section prescribing procedures for requiring energy consumption labeling of covered products [32].

4. A section providing for the establishment of targets for improved energy efficiency for covered products, to be followed, if deemed necessary, by energy efficiency standards for those products [33].

5. A section detailing the extent to which the above requirements preempt state regulation [34]. This section is reproduced in the appendix.

Consumer products covered by this part of the act include:

(1) Refrigerators and refrigerator-freezers.
(2) Freezers.
(3) Dishwashers.
(4) Clothes dryers.
(5) Water heaters.
(6) *Room air conditioners.*
(7) *Home heating equipment, not including furnaces.*
(8) Television sets.
(9) Kitchen ranges and ovens.
(10) Clothes washers.
(11) Humidifiers and dehumidifiers.
(12) *Central air conditioners.*
(13) *Furnaces.*
(14) Any other type of consumer product which the Administrator of the Federal Energy Administration classifies as a covered product . . . [35].

These products are primarily consumer products, and not generally the type used in commercial and industrial operations. Therefore, for purposes of this discussion, the problem of preemption is limited to those listed products that might be used in commercial or industrial operations and to other such products that the administrator might classify as covered. Listed products in this category include those used for heating and cooling (italicized in the list).

Scanning the preemption section, which is reproduced in the appendix, one notes that it, like the preemption section regarding automobile efficiency, provides an exemption for state government procurement [36]. The section provides that state "procurement standards" that "are more stringent than the corresponding Federal standards" are not superseded [37]. The question is whether state contract clauses requiring government contractors to use heating and cooling devices with prescribed efficiencies fall within the scope of "procurement standards." Were this the case, preemption of these clauses could be avoided merely by making their standards "more stringent than the corresponding Federal standards." A limited argument for this position can be based upon the narrow scope of the procurement exemption regarding automobiles; it is restricted, for example to ". . . automobiles procured for its [the state's] own use" [38]. It can be argued that since the Congress used very narrow language in the automobile part, its use of general language ("procurement standards") in the consumer product part implies a broad enough scope to include contractor compliance clauses. While worth making, this argument is not a sure winner. Accordingly, the discussion will continue on the assumption that contractor compliance clauses are not "procurement standards," within the meaning of the act.

Although the efficiency of heating and cooling devices used by government contractors is thus apparently beyond the realm of state or local controls [39], there is a loophole! Under 42 U.S.C. § 327 (the preemption section) state governments may avoid having a regulation superseded by demonstrating to the administrator of the Federal Energy Administration (FEA) that:

(A) there is a substantial State or local need which is sufficient to justify such State regulation;
(B) such State regulation does not unduly burden interstate commerce; and
(C) if there is a Federal energy efficiency standard applicable to such product, such State regulation contains a more stringent energy efficiency standard than the corresponding Federal standard [40].

Therefore, states desiring to prescribe the efficiency of their contractors' heating and cooling devices are advised to seek an FEA ruling to shield their efforts from federal preemption. Whether the implementation of this strategy constitutes a "substantial . . . need" is questionable, but some cases may present more compelling arguments than others.

What Should Contract Clauses Require?

As stated earlier, it is possible to require government contractors to adopt a wide variety of energy conservation measures. The federal discussion of preemption demonstrates that requirements by states regarding motor vehicles and heating and cooling devices must be carefully considered before implementation. In addition, it is necessary, or at least wise, to limit the scope and the duration of the requirements. Scope should be limited to the practices and equipment that the contractor uses in performing the government contract; duration should be limited to the life of the contract. These limitations will make enforcement of the requirements much simpler, and reduce the burden on the contractor. Reducing the burden of compliance will enhance competition for government contracts.

Within these constraints, state and local governments may wish to consider implementing requirements set out below. Implementation could be by either statute or executive order, although a statutory approach may be more clear-cut. The list below is by no means comprehensive, and any energy conservation measure that can fit within the above constraints is a likely candidate.

1. Requirements for a certain level of energy efficiency in lighting of work areas used in performance of the contract [41].
2. Requirements for fuel efficiency of motor vehicles used in performance of the contract, to the extent that motor vehicles (e.g., certain trucks) are not covered by federal fuel economy standards [42].
3. Requirements that contractors provide employees working on the contract with paratransit, i.e., vans for vanpooling [43].
4. Requirements that work schedules under the contract be staggered to alleviate traffic congestion in the vicinity of the workplace during rush hours [44].
5. Requirements that contractors limit use of employee parking facilities to carpools (minimum of four riders per car) [45].

Discussion of the conservation potential of these measures is beyond the scope of this book; footnotes provide references to more detailed treatment. In particular, readers are referred to three other books in this series: *Building to Save Energy: Legal & Regulatory Approaches*, by Grant P. Thompson; *Energy Efficiency in Industry: A Guide to Legal Barriers and Opportunities*, by Norman L. Dean; and *Saving Energy in Urban Transportation*, by Durwood Zaelke.

If the contract clause strategy is to be effective, compliance with its requirements must be inexpensive. Expensive or otherwise burdensome requirements will tend to limit the number of bidders for government contracts, and the reduction in competition may be costly to the government. In addition, governments should remember that the contract price will reflect the cost of complying with the energy conservation requirements. In light of this, every effort should be made to limit the cost of compliance and to publicize the energy cost savings that will accrue to contractors that comply. To emphasize savings, contractors might be required to keep records of their energy use, and its cost, before and after they begin observing the conservation measures.

In closing, the author wishes to stress that the strategies presented in this chapter are cumbersome from the standpoint of administrative detail—red tape—for both the purchasing agency and the government contractor. These strategies, and any alternatives that might prove simpler to implement, should be carefully examined before any steps are taken to put them into effect.

NOTES TO CHAPTER 3

1. Council of State Governments, *State and Local Government Purchasing* (Lexington, Kentucky, March 1975), pp. A.1—A.13.

2. *Ibid.*, p. 5.1. *See generally*, *Ibid.*, pp. 5.1—5.5 for discussion of bidder prequalification and the "Bidder's List."

3. *Ibid.*, p. 5.2.

4. *Ibid.*

5. *Ibid.*

6. Pub. L. No. 91—604, 84 Stat. 1676, 42 U.S.C. §§1857 et seq. (1970), ELR 41201.

7. Pub. L. No. 92—500, 86 Stat. 81b, 33 U.S.C. §§1251 et seq. (Supp. 1973), ELR 41101.

8. Clear Air Amendments of 1970 § 306, 42 U.S.C. §1857h-4 (1970), ELR 41225; Federal Water Pollution Control Act Amendments of 1972 §508, 33 U.S.C. §1368 (Supp. 1973), ELR 41125. For a discussion of the procurement aspect of enforcement of these acts, *see* Ralph C. Nash, Jr., and Susan Linden, "Federal Procurement and the Environment," in *Federal Environmental Law*, ed. Erica L. Dolgin and Thomas P. Guilbert (St. Paul, Minnesota: West, 1974), pp. 460—72.

Note that on December 23, 1976, the United States Environmental Protection Agency made use of these provisions by listing, in the Federal Register, several U.S. corporations. That Federal Register entry (41 Fed. Reg. 55931—32) is reproduced in the appendix to this chapter, p. 101. See also, United States v. Del Monte, 9 ERC 1495 (D.P.R. 1976).

9. For penalty provisions *see* the Clean Air Amendments §113, 42 U.S.C. §1857c-8, ELR 41209; and the Federal Water Pollution Control Act Amendments of 1972 § 309, 33 U.S.C. §1319 (Supp. 1973), ELR 41115.

10. 42 U.S.C. §1857h-4(a), and 33 U.S.C. §1368(a).

11. *See generally*, Norman L. Dean, *The Environmental Law Institute State and Local Energy Conservation Project, Energy Efficiency in Industry: A Guide to Legal Barriers and Opportunities* (Cambridge, Massachusetts: Ballinger, 1977).

12. Pub. L. No. 94—163, 42 U.S.C. §§6201 et seq. (1975).

13. 42 U.S.C.A. §§6321—26.

14. 42 U.S.C.A. §6322(c).

15. 42 U.S.C.A. §6322(d).

16. *See* 40 C.F.R. Part 15, "Administration of the Clean Air Act and the Federal Water Pollution Control Act with Respect to Federal Contracts, Grants or Loans," and *Federal Environmental Law, supra* note 8 at 471.

17. "The Buy American Act," 41 U.S.C.A. §10(b). Note that this type of provision, since having an effect upon foreign trade policy, has been held to be solely the province of the federal government. *See* Council of State Governments, *State and Local Government Purchasing* (Lexington, Kentucky, 1975) pp. 9.4–9.5. *See also,* Commission on Government Procurement, *Report of the Commission on Government Procurement,* vol. 1 (Washington, D.C., 1972), and American Institute for Imported Steel, Inc. v. Office of General Services, 365 N.Y.S. 2d 56 (1975) (held that paragraph of general specifications in public contracts incorporating "Buy American" policy invalid under N.Y. State Finance Law §174); Bethlehem Steel Corp. v. Board of Commissioners of Department of Water and Power, City of Los Angeles, 80 Cal. Reptr. 800, 276 C.A.2d 221 (1969) (held California's "Buy American" policy unconstitutional); and 53 Ops. Atty. Gen. 72, 2–11–70 (Calif.) (held California's "Buy California" materials policy unconstitutional).

18. Exec. Order No. 11245, Exec. Order No. 11375. *See,* e.g., State of N.Y., *Official Compilation of Codes, Rules and Regulations,* Ch. XIX, Part 900.1, "N.Y. Educational Construction Fund," for sample state public contract clause prohibiting discrimination authorized under N.Y. Education Law §454, subd. 5 (McKinney 1969).

19. Walsh-Healey Act, 41 U.S.C.A. § 35.

20. Vietnam Veterans Readjustment Act, Pub. L. No. 92–540, 38 U.S.C. §1502 et seq.

21. For further discussion of public policy covenants in government contracts, see John W. Whelan and Robert S. Pasley, *Federal Government Contracts* (Mineola, New York: Foundation Press, 1975), pp. 646–762; and *Report of the Commission on Government Procurement, supra* note 17 at 111–24.

22. Dean, *supra* note 11.

23. Northern States Power Co. v. Minnesota, 447 F.2d 1143 (8th Cir., 1971).

24. Energy Policy and Conservation Act (EPCA) § 327, Pub. L. No. 94–163 (1975), 42 U.S.C.A. §6297; Motor Vehicle Information and Cost Savings Act *as amended* by Energy Policy and Conservation Act § 301, 15 U.S.C.A. §2009.

25. 42 U.S.C.A. §6297; 15 U.S.C.A. §2009.

26. The Motor Vehicle Information and Cost Savings Act defines them as follows:

Sec. 501. For purposes of this part:

(1) The term 'automobile' means any 4-wheeled vehicle propelled by fuel which is manufactured primarily for use on public streets, roads and highways: (except any vehicle operated exclusively on a rail or rails) and

(A) which is rated at 6,000 lbs. gross vehicle weight or less, or

(B) which—

(i) is rated at more than 6,000 lbs. gross vehicle weight but less than 10,000 lbs. gross vehicle weight

(ii) is a type of vehicle for which the Secretary determines by rule, average fuel economy standards under this part are feasible and . . .

(iii) is a type of vehicle for which the Secretary determines by rule, average fuel economy standards will result in significant energy conservation or is a type of vehicle which the Secretary determines is substantially used for the same purposes as vehicles described in subparagraph (A) of this paragraph. The Secretary may prescribe such rules as may be necessary to implement this paragraph. 15 U.S.C.A. § 2001.

27. 15 U.S.C.A. § 2009(a) (emphasis added).

28. 15 U.S.C.A. § 2009(c).

29. Energy Policy & Conservation Act (EPCA) §§ 321–339, 42 U.S.C. §§ 6291–6309.

30. EPCA § 322, 42 U.S.C.A. § 6292.

31. EPCA § 323, 42 U.S.C. § 6293.

32. EPCA § 324, 42 U.S.C. § 6294.

33. EPCA § 325, 42 U.S.C. § 6295.

34. EPCA § 327, 42 U.S.C. § 6297.

35. EPCA § 322, 42 U.S.C. § 6292 (emphasis added).

36. *See* EPCA § 327(c), 42 U.S.C. § 6297(c).

37. *Ibid.*

38. EPCA § 301, 15 U.S.C. § 2009.

39. EPCA § 327(a), 42 U.S.C. § 6297(a).

40. EPCA § 327(b) (2), 42 U.S.C. § 6297(b) (2).

41. *See* Grant P. Thompson, *The Environmental Law Institute State and Local Energy Conservation Project, Building to Save Energy: Legal and Regulatory Approaches* (Cambridge, Massachusetts: Ballinger, 1977). Note also that state plans under EPCA must include mandatory lighting efficiency standards for public buildings. 42 U.S.C. § 6322.

42. *See* Durwood J. Zaelke, *The Environmental Law Institute State and Local Energy Conservation Project, Saving Energy in Urban Transportation* (Cambridge, Massachusetts: Ballinger, 1977). (Chapter on paratransit.)

43. *See* Chapter 7 of this book, discussing management and maintenance strategies.

44. *See* Zaelke, *supra* note 42.

45. *Ibid.*, and Chapter 7 of this book.

Appendices to Chapter 3

THE UNITED STATES ENVIRONMENTAL PROTECTION AGENCY'S
DEBARMENT OF SEVERAL CORPORATIONS FROM PARTICIPATION
IN GOVERNMENT CONTRACTS

Environmental Protection Agency
[FRL 661–7]

List of Violating Facilities

Pursuant to Section 306 of the Clean Air Act (42 U.S.C. 1857 et seq., as amended by Public Law 91–604), the Federal Water Pollution Control Act, (33 U.S.C. 1251 et seq., as amended by Public Law 92–500), and Executive Order 11738, EPA has been authorized to provide certain prohibitions and requirements concerning the administration of the Clean Air Act and Federal Water Pollution Control Act with respect to Federal contracts, grants or loans. On April 16, 1975, regulations implementing the requirements of the statutes and the Executive Order were promulgated in the Federal Register (see 40 CFR Part 15, 40 F.R. 17124, April 16, 1975). Section 15.20 of the regulations provides for the establishment of a List of Violating Facilities which will reflect those facilities ineligible for use in nonexempt Federal contracts, subcontracts, grants, subgrants, loans, or subloans.

The representatives of any facility under consideration for listing are afforded the opportunity to appear at a Listing Proceeding conducted by the Director, Office of Federal Activities. Listing occurs when the Director determines there is adequate evidence of noncompliance with clean air or water standards. Federal, State, and local criminal convictions, civil adjudications, and administrative findings of noncompliance may serve as a basis for consideration of listing. However, in the case of a State or local civil adjudication or administrative finding, EPA may consider listing only at the request of the Governor.

The List of Violating Facilities is contained in two sublists. Sublist 1 includes those facilities listed on the basis of a conviction under section 113(c) of the Clean Air Act or section 309(c) of the Federal Water Pollution Control Act. Sublist 2 includes those facilities listed on the basis of any injunction, order, judgment, decree or other form of civil ruling by a Federal, State or local court issued as a result of noncompliance; or a conviction in a State or local court for noncompliance; or on the basis of noncompliance with an order under section 113(a) of the Clean Air Act or section 309(a) of the Federal Water Pollution Control Act, or having been subjected to equivalent State or local proceedings to enforce clean air or water standards.

The purpose of this Notice is to add the Ashland, Kentucky, facility of the Allied Chemical Corporation, Semet-Solvay Division, to Sublist 1 of the List of Violating Facilities.

No agency in the Executive Branch of Government shall enter into, renew, or extend any nonexempt contract, subcontract, grant, subgrant, loan or subloan where a facility listed would be utilized for the purpose of any such agreement.

Pursuant to this authority, the Director, Office of Federal Activities, U.S. Environmental Protection Agency, certifies that the following facilities have been placed on the List of Violating Facilities as of December 17, 1976. The List of Violating Facilities will be revised periodically as any listings or de-listings occur.

List of Violating Facilities

Sublist 1

Allied Chemical Corporation, Semet-Solvay Division, Ashland, Kentucky

Sublist 2

Del Monte de Puerto Rico, Inc., Mayaguez, Puerto Rico
Star-Kist Caribe, Inc., Mayaguez, Puerto Rico

Dated: December 17, 1976.

Rebecca W. Hanmer,
Director,
Office of Federal Activities.

[FR Doc. 76–37749 Filed 12–22–76; 8:45 am]

Source: 41 Fed. Reg. 55931–32 (December 23, 1976).

ERDA PRESS RELEASE*

A portent of things to come in the analysis of energy used in the performance of contracts

No.   75–264   FOR IMMEDIATE RELEASE
Contact: Robert J. Griffin   (Wednesday, December 24, 1975)
Tel.   202/376–4064

ERDA AWARDS CONTRACT TO DETERMINE
ENERGY USE IN THE BUILDING PROCESS

The Energy Research and Development Administration (ERDA) has awarded a $99,463 contract to the University of Illinois to analyze the energy used in the construction of buildings from raw materials to finished structure.

Architect, Richard G. Stein, FAIA, of New York City and Professor Bruce M. Hannon of the University's Center for Advanced Computation will direct the project.

Austin N. Heller, ERDA Assistant Administrator for Conservation, who announced the contract, said "By tracking the energy consumed in all phases of building construction, and in the processing of building materials, information will be developed which design professionals, contractors and building owners can use to select energy efficient techniques and materials. This research project will also aid in planning future research and development programs to conserve energy in all types of buildings."

Mr. Stein and Professor Hannon have had considerable experience in this field. Mr. Stein has earned a national reputation as an innovative architect in the field of energy conservation, and Professor Hannon has been instrumental in applying computer technology to broad-scale analysis of industry.

Up to now, little attention has been paid to conserving the nation's energy resources in the many separate steps of construction, starting from the processing of raw materials to the finishing operations of building construction. The nature of the building industry, which has been traditionally fragmented among tens of thousands of separate producers of building materials, contractors, and sub-contractors, has made it difficult to examine energy utilization from a more comprehensive viewpoint.

*Source: United States Energy Research and Development Administration (ERDA), *Information from ERDA*, vol. 1, no. 39 (Washington, D.C., Dec. 24, 1975), p. 3.

EXCERPT FROM THE ENERGY POLICY AND CONSERVATION ACT, 42 U.S.C. §6201 ET SEQ. (1975). THE FOLLOWING SECTION DEALS WITH PREEMPTION OF STATE REGULATION OF CERTAIN ASPECTS OF ENERGY EFFICIENCY OF CERTAIN OTHER CONSUMER PRODUCTS

§6297. State Laws; Warranties

(a) This part supersedes any State regulation insofar as such State regulation may now or hereafter provide for—

    (1) the disclosure of information with respect to any measure of energy consumption of any covered product—

        (A) if there is any rule under section 6293 of this title applicable to such covered product, and such State regulation requires testing in any manner other than that prescribed in such rule under section 6293 of this title, or

        (B) if there is a rule under section 6294 of this title applicable to such covered product and such State regulation requires disclosure of information other than information disclosed in accordance with such rule under section 6294 of this title; or

    (2) any energy efficiency standard or similar requirement with respect to energy efficiency or energy use of a covered product—

        (A) if there is a standard under section 6295 of this title applicable to such product, and such State regulation is not identical to such standard, or

        (B) if there is a rule under section 6293 or 6294 of this title applicable to such product and such State regulation requires testing in accordance with test procedures which are not identical to the test procedures specified in such rule.

(b) (1) If a State regulation provides for an energy efficiency standard or similar requirement respecting energy use or energy efficiency of a covered product and if such State regulation is not superseded by subsection (a) (2) of this section, then any person subject to such State regulation may petition the Administrator for the prescription of a rule under this subsection which supersedes such State regulation in whole or in part. The Administrator shall, within 6 months after the date such a petition is filed, either deny such petition or prescribe a rule under this subsection superseding such State regulation. The Administrator shall issue such a rule with respect to a State regulation if and only if the petitioner demonstrates to the satisfaction of the Administrator that—

        (A) there is no significant State or local interest sufficient to justify such State regulation; and

        (B) such State regulation unduly burdens interstate commerce.

    (2) Notwithstanding the provisions of subsection (a) of this section, any State regulation which provides for an energy efficiency standard or similar requirement respecting energy use or energy efficiency of a covered product shall not be superseded by subsection (a) of this section if the State prescribing such standard demonstrates and the Administrator finds, by rule, that—

(A) there is a substantial State or local need which is sufficient to justify such State regulation;

(B) such State regulation does not unduly burden interstate commerce; and

(C) if there is a Federal energy efficiency standard applicable to such product, such State regulation contains a more stringent energy efficiency standard than the corresponding Federal standard.

(c) Notwithstanding the provisions of subsection (a) of this section, any State regulation which sets forth procurement standards for a State (or political subdivision thereof) shall not be superseded by the provisions of this part if such State standards are more stringent than the corresponding Federal standards.

(d) For purposes of this section, the term "State regulation" means a law or regulation of a State or political subdivision thereof.

(e) Any disclosure with respect to energy use, energy efficiency, or estimated annual operating cost, which is required to be made under the provisions of this part, shall not create an express or implied warranty under State or Federal law that such energy efficiency will be achieved, or that such energy use or estimated annual operating cost will not be exceeded, under conditions of actual use.

✳ *Chapter 4*

Energy Impact Assessment

Twenty-five states, the Commonwealth of Puerto Rico, and the federal government have enacted statutes requiring that government agencies prepare environmental impact statements or "reports" prior to the undertaking of certain types of government action affecting the environment [1]. These statements are intended to disclose the full extent of environmental damage that may result from a particular government action [2]. In addition, the federal requirements and most state requirements call for discussion of alternatives to the proposed action [3]. This mechanism stands as the major watchdog of the nation's environment.

While there is an increasing awareness that the production and consumption of energy contribute substantially to environmental degradation, the legislation noted above does not contain a commensurate requirement for analysis of the potential energy impact. Only ten of the twenty-five governments requiring impact statements expressly provide for analysis of energy impact, and the analysis so required is limited in scope [4]. The low level of consideration is unfortunate, for an environmental impact analysis is fundamentally incomplete without an evaluation of energy impact.

To the extent that the types of government actions that require impact statements constitute "public works," they are appropriate for consideration in this book. An analysis of the energy consumption that is attributable, both directly and indirectly, to the construction and operation of a proposed major government facility can contribute greatly to the ultimate reduction of that consumption. Once analyzed, a project's energy effects may demonstrate the wisdom of substantially altering or totally abandoning the project.

WHAT IMPACTS?

Chapters 1 through 3 have discussed strategies which focus on a wide variety of energy uses, both direct and indirect. Readers are now aware that energy use is triggered by many mechansims besides light switches and gas pedals. Direct energy use is easy to spot; it usually produces heating, cooling, lighting, or motion. Indirect energy use is a measure of past, and sometimes future, direct energy use surrounding the creation of a product. It includes the energy required to create the product, transport it to its place of utilization, and remove and dispose of it when it is no longer useful. Since the energy is expended to bring a product to a useful form and to place it in the hands of the user, it seems reasonable to say that the user of the product is also the consumer of this energy.

As noted in Chapter 2, past and future energy use recorded in a particular product may include energy consumed (1) in extracting or harvesting component raw materials; (2) in refining those materials; (3) in fabricating the finished item; (4) in transporting the components and the finished item to the consumer; and (5) in disposing of the item after use [5].

An evaluation of the energy needed to construct and operate a public works project requires analysis of a myriad of direct and indirect uses. It thus might include energy used to (1) heat, cool, and light offices of those planning and designing the project; (2) transport the land surveyors around the site; (3) excavate the land in preparation for construction; (4) extract, refine, and fabricate the construction materials; (5) transport materials to the site; (6) power the machines used to construct the project; (7) heat, light, and otherwise operate the facility once it is constructed; (8) extract, refine, and fabricate the replacement parts needed to keep the facility in operation; (9) demolish the facility at the end of its use and prepare the site for another use; and (10) transport any salvageable materials to places of alternative use [6]. Furthermore, since the energy impact extends far beyond the boundaries of the construction site, it may be appropriate to also consider: (1) a portion of the energy used to produce the construction tools; (2) fuel used by construction employees in traveling to work; (3) fuel used by facility users to travel to the facility; (4) a portion of the energy used to construct the offices of project planners; (5) a portion of the energy used to construct the demolition equipment; (6) a portion of the energy used to construct the transport vehicles used at various points in the process; etc. The chain seems endless. Even the energy used to produce and transport the *energy* that is used in all the other processes must be considered.

A recent article by Michael Gerrard, *Disclosure of Hidden Energy Demands: A New Challenge for NEPA*, provides a very good discussion of this problem [7].

EXISTING APPROACHES

As mentioned in this chapter's introduction, several states have already developed limited requirements for energy impact analysis. California recently amended its Environmental Impact Report legislation, applicable to all state and local government activities affecting the environment and any private activity subject to public licensing [8], to require a detailed statement of "mitigation measures proposed to minimize the impact, including but not limited to measures to reduce wasteful, inefficient and unnecessary consumption of energy" [9]. Similarly, environmental impact statements prepared in New York must include a discussion of the project's effect on "use and conservation of energy resources" [10]. Several other states have developed guidelines under existing legislation, which call for a discussion of energy impacts as part of their environmental impact reporting requirements [11].

In addition, several states have proposed legislation specifically requiring energy impact statements prior to construction of public buildings [12]. These proposals outline the energy effects that should be discussed in public works-related energy impact analyses. Required discussion topics include the types and quantities of energy resources necessary for operation and maintenance of the completed project; heating and cooling requirements, including methods of insulation and ventilation rates; energy storage systems; and analysis of comparative costs of practicable alternative energy sources and utilization measures [13].

HOW TO ASSESS INDIRECT
ENERGY IMPACTS

Existing approaches tend to focus upon the energy impacts of public facilities once their construction is completed. While these impacts are highly significant, failure to include an analysis of the energy required to produce and assemble the materials to construct the facilities results in overlooking the substantial, near term energy cost incurred before the facilities are ever used. There is, however, a reason for the current tendency to limit energy analysis to direct facility consumption—quantification of indirect consumption can be both complex and confusing.

How on earth is the impact statement preparer to determine how many watt-hours of electricity were used to power the impact wrench that tightened the 363rd bolt in the truck that was used to haul 1/3,000th of the insulation required to construct the sports arena? And given the total useful life of the truck, how much of the bolt-tightening energy should be allocated to this one trip? These are among the millions of questions that would plague researchers trying to trace all the steps leading to the construction of a public facility. Obviously, such an analysis would outstrip the construction of the Pyramids in its expense and difficulty.

While this book does not pretend to be a manual for the preparation of impact statements, it will describe one technique for estimating indirect energy use with relative simplicity. The method is input-output analysis as applied by Bruce Hannon, Robert A. Herendeen, and Clark W. Bullard, III, of the Center for Advanced Computation at the University of Illinois to determine the energy cost of goods and services [14]. The result is a coefficient of energy intensity for each type of product or activity. The coefficient tells how many BTU are required to produce one dollar's worth of the product or activity. The figure is obviously a very rough average, but it can be valuable for means of comparison, and its use is substantially cheaper than trying to trace every material through every stage of its production.

The willing impact statement preparer is thus presented with a number for each of 357 types of goods and services [15]. The number expresses BTU per dollar [16]. For example, according to the 1967 table prepared by Center for Advanced Computation (CAC), the number for ice cream (if produced by coal) is 20,321, which means that 20,321 BTU were required for producing one dollar's worth of ice cream in 1967 [17]. Accordingly, the impact statement preparer need only identify the types of goods and services involved in the project under evaluation, determine the dollar cost of the goods and services, and multiply the dollar cost by the appropriate numbers from the CAC table. The people from CAC have done all the rest. Their numbers are based on analysis of transactions among the sectors producing the 357 types of goods and services, and represent crude nationwide averages of energy required to produce each type [18]. Calculations include analysis of feedback loops, as where producers of rayon cord for tires use similar tires on their own business automobiles [19].

Application of the CAC method requires stern caveats. The method has many flaws, and must be applied within the context of a framework that can compensate for at least some of them. The Michael Gerrard article mentioned above contains an example of such a framework. Some of the major problems are outlined below.

1. Since a number of years are required to prepare the coefficient, the data on which it is based are inherently very old, and presently available data antedate the "energy crisis."
2. The coefficients represent a national average, and ignore variance in energy efficiency among individual members (manufacturers) of a single sector.
3. The data from which the coefficients are derived are transmitted by industry to the United States Department of Commerce. The primary use for which the data are intended has little to do with energy, and the data may not be sufficiently precise or well-suited for energy analysis.
4. The method does not, and by definition cannot, take new technology into account [20].

The purpose here in presenting this use of input-output (I/O) analysis is merely to encourage the reader to examine further the value of analysis of indirect energy use. Any attempt to apply I/O should be preceded by substantial additional study.

On the positive side, however, I/O analysis is sufficiently workable to be the method recommended by the Stanford Research Institute for estimating energy savings for the procurement programs under the state energy conservation plans of the Energy Policy and Conservation Act [21].

IMPLEMENTATION OF ENERGY IMPACT ANALYSIS REQUIREMENT

For States With Existing Environmental Impact Statement Requirements

Given the pervasive effects of energy production and consumption upon the environment, those state agencies that are responsible for environmental impact statement preparation could easily justify undertaking energy impact analysis. Given, however, that such analysis may constitute a substantial burden, these agencies may be reluctant to volunteer for the task. In such cases, it is a simple matter for the legislature (or the executive) to amend the impact statement requirements so as to include analysis of both direct and indirect energy consumption.

For States With No Environmental Impact Statement Requirements

Where the state has no requirement for an environmental statement, it is up to the legislature to pass legislation requiring energy impact analysis. While it is undesirable ultimately to require separate energy and environmental impact statements, energy analysis requirements should be considered as a forerunner to requiring analysis of the full range of environmental effects of a public works project.

FACILITY LOCATION: A PLANNING DECISION MERITING SPECIAL CONSIDERATION

The location of government buildings and other public works has a substantial effect upon the amount of energy used for transportation of employees and other users of the public facilities. While it is expected that this consideration will be included in the energy impact analyses described above, location is of such singular importance as to require specific, if very brief, treatment.

Three aspects of location are of primary importance to energy use: (1) proximity of public facilities to residential areas; (2) proximity of public facilities to each other and to private facilities providing consumer services; (3) proximity of facilities to mass transit [22]. These aspects affect the distances that government employees and other building users must travel for employment, shopping, other services, and recreation, as well as the efficiency of their transport. By reducing the distances and improving the efficiency, officials responsible for siting decisions can foster substantial energy saving [23].

This discussion is not intended to provide operational land use strategies, but rather to sensitize the reader to the importance of siting decisions. Land use and energy consumption are, however, discussed in considerable depth in another book in this series, Corbin Crews Harwood's *Using Land to Save Energy* [24]. In addition, transportation is discussed in Durwood Zaelke's *Saving Energy in Urban Transportation*.

Proximity to Residential Areas

Isolating government facilities in areas remote from the residences of the potential users is obviously energy-wasteful. This is not necessarily to say that the garbage incinerator should be in the town square, but that facilities that are compatible should be integrated with residential areas [25]. Likely candidates include office buildings, libraries, hospitals, and performing arts centers.

A probable major impediment to this siting policy is the higher land prices in residential as opposed to outlying areas. Decisionmakers faced with the increased cost may have difficulty in justifying it, because the added expenditure will not save government money. On the other hand, since such facilities are often constructed for the benefit of the community, it may be argued that this siting policy will reduce transportation costs for members of the community. It is probable that the policy will also reduce air pollution from motor vehicle exhaust.

Proximity to Other Government and Private Facilities

If facilities, both public and private, such as office buildings, stores, tonsorial parlors, theatres, hospitals, doctors' offices, libraries, civic centers, swimming pools, restaurants, etc., are located close to each other, workers and consumers can fulfill many needs with one trip. Vehicle miles traveled are cut, and energy is saved.

One impediment to the policy of clustering facilities may be a government's political motivation to spread its facilities among neighborhoods, so that no one area feels deprived or burdened. It should be possible to explain that clustering provides better service even to residents of areas distant from the cluster. It is easier to travel to one location, even if somewhat distant, than to travel to every neighborhood, both near and far.

Proximity to Mass Transit [26]

If government office buildings are located close to mass transit, government employees are likely to use this transit. Furthermore, convenient transit facilitates elimination of parking lots or stringent parking management programs [27]. Providing employees with an attractive alternative transport mode should take the sting out of denying them parking spaces.

Land in the vicinity of mass transit stations tends to appreciate substantially around the time operation begins. In general, people like to live reasonably near stations, and businesses find it profitable to locate where customers can conveniently reach them. Consequently, the high price of land near stations may prevent optimal locating of government buildings. One way of circumventing the problem would be to plan the buildings at the same time as the mass transit, and acquire land for both before it had appreciated in value. This may require state and federal coordination.

The Siting Decision

Generally, officials charged with choosing sites for public works are accorded broad discretion [28], especially in choosing sites for office buildings [29]. Broad powers of discretion should greatly facilitate the implementation of the policies outlined in this chapter.

Twenty-five states now have some form of environmental impact statement requirement, which is primarily applicable to public works [30]. The policies espoused in this chapter are precisely the types of considerations that the impact statement process is designed to encourage. State Environmental Impact Statements (EISs) should therefore serve as vehicles, rather than impediments, to strategy implementation.

Note, finally, that the effects discussed above are often multiplied, due to the stimulating effect that public works projects may have upon private development [31].

NOTES TO CHAPTER 4

1. National Environmental Policy Act of 1969, 42 U.S.C. §§4321–47 (1970), *as amended* by Pub. L. No. 94–83 (1975); California Environmental Policy Act of 1970, Cal. Pub. Res. Code §§21000–21174 (West 1970), *as amended* (Supp. 1976); Connecticut Environmental Policy Act of 1973, Conn. Gen. Stat. Ann. ch. 439, §§22a–1 et seq. (Supp. 1974–75); Hawaii Rev. Stat. ch. 341 (1974); Ind. Stat. Ann. §35–5301 et seq. (Supp. 1971); Maryland Environmental Policy Act of 1973, Md. Ann. Code Art. 41, §§447–451 (Supp. 1973) and Md. Ann. Code, Natural Resources, §§1–301 et seq. (1974), *as amended* (Supp. 1975); Ann. Laws of Mass. ch. 30, §§61–62 (1973), *as amended* (Supp. 1975); New York State Environmental Quality Review Act of 1975, N.Y. Environmental Conservation Law §8–0101 et seq. (McKinney Supp. 1975–1976); North Carolina Environmental Policy Act of 1971, N.C. Gen. Stat. ch. 113A (Supp. 1973); Minnesota Environmental Policy Act of 1973, Minn. Stat. Ann. ch. 116D (Supp. 1976); Montana Environmental Policy Act of 1971, Rev. Code of Mont. §§69–6501 et seq. (Supp. 1973), *as amended* (Supp. 1975); South Dakota Environmental Policy Act, S.D. Comp. Laws 1967 ch. 11–1A (Supp. 1976); Virginia Environmental Policy Act of 1973, Va. Code Ann. §10–177 et seq. (Supp. 1976), *as amended* by Va. Code Ann. §§2.1–51.9, §10.181, §10.183 and §10.185 (Supp. 1976); State Environmental Policy Act of 1971, Rev. Code Wash. ch. 43.21C (Supp. 1976); Wisconsin Environmental Policy Act of 1971, Wisc. Stat. Ann. ch. 1, §1.11 et seq. (Supp. 1975–76); Puerto Rico Environmental Policy Act, 12 Laws P.R. Ann. §1121 et seq. (1970); Mich. Exec. Order No. 1971–10, as superseded by Mich. Exec. Order No. 1974–4 (May 1974); N.J. Exec. Order No. 53 (October 15, 1973); "Policy for the Environment," (a Voluntary Memorandum of Understanding signed by all Texas state agencies) (adopted January 1, 1973); Game and Fish Commission Policy of July 2, 1971, Memorandum by the Arizona Game and Fish Commission, "Requirements for Environmental Impact Statements," issued June 9, 1971; Delaware Coastal Zone Act, 7 Del. Code Ann. tit. 7, §§7001 et seq. (Rev. 1974); Delaware Wetlands Law of 1973, Del. Code Ann. tit. 7, §§6601 et seq. (Rev.

1974); Ga. Code Ann. ch. 95A—1, §1241(e) (1) (1975); Nebraska Department of Roads Action Plan (1973) (being rewritten to meet new directives of the U.S. Department of Transportation: FHPM 771 and 772); Nev. Rev. Stat. ch. 704.820 et seq. (1973); Coastal Area Facility Review Act, N.J.S.A. 13:19—1 et seq. (Supp. 1974—75); New Jersey Wetlands Act of 1970, N.J.S.A. 13:19—1 et seq. (Supp. 1974—75); Miss. Code Ann. §49—27—11(i) (Supp. 1974); N.D. Cent. Code §54—01—05.04 (1974); Utah, Exec. Order of Governor (August 27, 1974).

2. *See generally*, Environmental Law Institute, *Federal Environmental Law*, ed. Erica L. Dolgin and Thomas P. Guilbert (St. Paul, Minnesota: West 1974), pp. 238—419; and Frederick R. Anderson, *NEPA in the Courts: A Legal Analysis of the National Environmental Policy Act* (Washington, D.C.: Resources for the Future, 1973).

3. *See also*, T.C. Trzyna, *Environmental Impact Requirements of the States: NEPA's Offspring 19022*, EPA—600—5—73—006 Office of Research and Development (Washington, D.C.: EPA, April 1974).

4. The governments requiring a discussion of energy impacts in EISs include the following: federal government, California, Delaware, Maryland, Massachusetts, Michigan, Montana, New Jersey, New York, and Wisconsin. *See also*, Michael Gerrard, "Disclosure of Hidden Energy Demands: A New Challenge for NEPA," *Envir. Affairs* 4 (1975): 664; and Norman Dean, *Energy Efficiency in Industry: A Guide to Legal Barriers and Opportunities* (Cambridge, Massachusetts: Ballinger, 1977).

5. *See* Clark W. Bullard, III, and Robert A. Herendeen, "Energy Impact of Consumption Decisions," *Proceedings of the IEEE* (Institute of Electrical and Electronics Engineers, Inc.), 63, no. 3 (March 1975): 484—93; Gerrard, *supra* note 4; and Denis Hayes, *Energy: The Case for Conservation* (Washington, D.C.: Worldwatch Institute, January 1976), pp. 60—65.

6. The reader may notice a similarity between the length of this list and the one on page 20, where life cycle costing is discussed. Energy impact analysis is much more complex than life cycle costing, however, since determining energy use at various steps in construction and operation is more difficult than determining the cost of the steps.

7. Gerrard, *supra* note 4.

8. *See*, Friends of Mammoth v. Mano County, 8 Cal. 3d 247, 4 ERC 1593, 2 ELR 20673 (Cal. Sup. Ct. 1972).

9. Cal. Pub. Res. Code §21100(c) (West Supp. 1976).

10. N.Y. Environmental Conservation Law §8.0109(2) (b) (McKinney 1975).

11. Those states are Maryland, Massachusetts, Michigan, Montana, New Jersey, and Wisconsin. *See also* Gerrard, *supra* note 4.

12. *See, e.g.*, State of Florida, Senate No. 1312 (Feb. 1975), House No. 1971 (Feb. 1975); State of Connecticut, Proposed Bill No. 7488 (Jan. 1975); State of Maryland, House Bill No. 46, art. 43 §448 (Jan. 1975).

13. *Ibid*.

14. *See* Robert A. Herendeen and Clark W. Bullard, III, *Energy Cost of Goods and Services*, 1963 and 1967 (Urbana; Center for Advanced Computation, University of Illinois, November 1974); and Bullard and Herendeen, *supra* note 5 at 484—93.

15. Herendeen and Bullard, *supra* note 14 at 1.

16. *Ibid.*, p. 24.

17. *Ibid.*, p. 32.

18. *Ibid.*

19. *Ibid.*, p. 1.

20. Conversation with Stephen O. Anderson, Sierra Club researcher and consultant to the Environmental Law Institute. *See also*, Gerrard, *supra* note 4; and *Draft Proceedings Report*, Net Energy Analysis Workshop, held at Stanford University, August 25–28, 1975.

21. *See* Edward Dickson and James Bick, *Methodology for Estimating Energy Savings for State Conservation Plans* (Washington, D.C.: Stanford Research Institute, April 1976).

22. Interview with Corbin Crews Harwood, Environmental Law Institute, May 1976.

23. A recent publication of the United States General Services Administration notes: "In our present life style, about three times as much energy is expended getting building occupants from home to work and back again than is spent in energy to service the occupants in the building." United States General Services Administration (GSA), *Energy Conservation Guidelines for Existing Office Buildings* (Washington, D.C., 1975), p. 4–1.

24. *See* Corbin Crews Harwood, *The Environmental Law Institute State and Local Energy Conservation Project, Using Land to Save Energy* (Cambridge, Massachusetts: Ballinger, 1977); and Public Building Service, U.S. General Services Administration, *Energy Conservation Design Guidelines for New Office Buildings*, 2nd ed. (Washington, D.C., July 1975), p. 4–1.

25. Hopefully, the garbage incinerator will burn only items that are not recyclable, and the burning will be used to produce electricity or industrial steam. *See* U.S. Environmental Protection Agency, *Third Report to Congress, Resource Recovery and Waste Reduction* (Washington, D.C.: GPO, 1975), pp. 33–44; and House Committee on Government Operations, Solid Waste—Materials and Energy Recovery, H.R. Rep. No. 94–1319, 94th Cong., 2d sess. (1976).

26. Note that of the three policies mentioned here, this is the only one mentioned in the U.S. General Services Administration's energy conservation handbook for new office buildings. *See Energy Conservation Design Guidelines for New Office Buildings*, *supra* note 2 at 4–1.

27. *See*, Durwood J. Zaelke, *The Environmental Law Institute State and Local Energy Conservation Project, Saving Energy in Urban Transportation* (Cambridge, Massachusetts: Ballinger, 1977). (Parking Management Strategies.)

28. *See* 26 Am. Jur. 2d *Eminent Domain* §§25–72; 26 Am. Jur. 2d *Public Works and Contracts* §22; Cincinnati v. Wegehoft, 119 Ohio St. 136, 162 N.E. 389 (1928).

29. *See* 26 Am. Jur. 2d *Eminent Domain* §40.

30. *See* Chapter 7 of this book.

31. *See*, The Council on Environmental Quality, *The Growth Shapers* (Washington, D.C.: GPO, 1976).

Chapter 5

Cooperative Purchasing: A Collateral Strategy

Many purchasing strategies share a common restraint: they require a disproportionate initial expense on the part of the implementing jurisdiction. This extra expense may be incurred in the form of increased purchase prices of supplies and equipment or in the administrative expense of strategy implementation or both. These extra costs are frequently recoverable "with interest" in the long term, as with life cycle costing, but the problem of higher initial cost remains. Since in this relatively difficult economic era most local and state governments are looking for cost savings, higher initial cost could sound the death knell of conservation strategies.

One very good way to minimize higher initial costs, and to save money even where energy may not be crucial, is to engage in *joint, cooperative purchasing*. There are several variations on the basic concept of buying a particular type of good for more than one jurisdiction under a single contract, but variations share common advantages. These basically result in reduced purchasing prices and reduced administrative costs—savings that directly parallel the types of increased cost that conservation strategies may engender. Reduced prices may result from improved specifications and from increased purchasing volume. Increased volume, i.e., a larger market, may also provide incentive for government contractors to develop innovative, energy-efficient technology. Certain administrative costs remain about the same regardless of the volume of the purchase. Spreading these costs over the budgets of several governments reduces each government's expenditure, leaving more funds available to implement energy conservation strategies [1].

In some measure, cooperative purchasing is in itself an energy conservation strategy, to the extent that reducing duplicative purchasing yields energy savings. Savings might occur, for example, in the consolidation of testing laboratories. This discussion of purchasing, however, casts it primarily in the role of a collateral strategy, which facilitates the implementation of conservation strategies already presented. The major thrust of this strategy is to remove barriers to cooperative efforts among state and local public purchasing agencies so that they may freely negotiate among themselves for the acquisition or use of supplies, services, and facilities.

Two underlying considerations merit preliminary attention. First, a decision must be made as to whether the program will be permissive or mandatory. The Council of State Governments favors a permissive approach, as does the American Bar Association's Coordinating Committee on a Model Procurement Code [2]. Since local units may see cooperative purchasing as a threat to local autonomy, making the program permissive encourages local acceptance [3]. Texas is the only state that has any kind of mandatory program, and it is limited to school districts receiving equalization aid from the state and merely requires that buses, tires, and tubes be purchased from the state board of control [4]. The second consideration is protecting open competition [5]. The principles of competitive bidding need to be adequately guarded by assuring that provisions governing bidding procedures and contract administration are uniform among the cooperating jurisdictions.

By the end of 1976, forty-three states had some form of cooperative purchasing program [6]. There are four basic approaches. The first approach permits complete freedom of contract between local governments and state contract vendors; the state furnishes only a listing of commodities, vendors, and prices [7]. A problem with this approach is that while it provides extreme flexibility, it leads to loss of contract control by the state, and in some instances may inhibit the obtaining of lower prices, e.g., where a local jurisdiction is closer to a substitute source of supply [8].

A second approach allows local units to purchase directly from state contract vendors but requires that local agencies send a copy of the purchase order to the state. This approach, used in Illinois, allows for slightly better control over the contracting process [9].

A third approach, and probably the preferred one, requires that all units be actual parties to the contract [10]. The participating units must agree to be bound contractually in the same manner as the state. California follows this strategy by requiring participating local units and school boards to estimate their purchasing needs in advance and submit standard requisition forms, which have been authorized by local governing boards, to the state [11]. Under such an arrangement, participating governments are bound contractually to order according to their requirements and to assume responsibility for delivery and payment.

A fourth method of cooperative purchasing is based on bulk purchasing and central warehousing, where the largest participating unit purchases and stores large quantities of commodities and other agencies order and pick up the goods [12]. This approach is most efficiently useful where the selected commodities can be delivered periodically and stored. Open competition is assured since bids are advertised for the definite amount required by all participating units [13].

Suggestions have been made regarding variations of these basic approaches and require consideration of (1) the kinds of commodities and services that could be jointly purchased and (2) methods of assuring quality. For example, New Jersey law permits local units to enter into cooperative contracts for (1) purchasing commodities such as fuel for heating and equipment; (2) providing services, such as snow plowing and removal, garbage collection and disposal, and recycling; (3) using data-processing equipment; (4) leasing or servicing of vehicles, equipment, and machinery; and (5) acquiring telephone services [14]. The Council of State Governments recommends pooling specialized technical functions such as commodity cataloging, specifications, inspection and testing, training, economic analysis, buying strategies, and disposing of surplus equipment as part of the cooperative purchasing program [15]. Implementation of this recommendation would also affect quality assurance.

Other methods are possible. One author has suggested a committee approach—commodity committees would be established, and each community would handle the entire operation in relation to the specific commodity, i.e., collect requisitions, develop specifications, etc. [16]. Other methods range from having each government unit do its own testing to the use of state facilities, joint use of private laboratory services, and inserting quality assurance provisions in bid proposals [17]. The ABA Model Procurement Code—Preliminary Working Paper No. 1 would establish a centralized state information system to disseminate information concerning the nature, cost, quality, and quantity of items [18]. Furthermore, states would be authorized to supply informational services, such as forms and manuals [19], and technical services such as testing and training methods and facilities [20] to local public purchasing agencies.

Cooperative purchasing programs are both valid and feasible means for governments to conserve time, money, and energy. Time savings result from the pooling of information, expertise, and services, which in turn reduce duplication of work and time spent developing specifications and testing products. The money savings, as stated earlier, result from the relative economy of purchasing commodities in larger volumes, as well as from elimination of administrative duplication. One survey found a typical dollar savings of 10—20 percent on most items [21]. As examples, in the first twenty-one months of its program, Colorado agencies saved $580,000; Illinois participants saved $838,000 over two years; and each New Hampshire participant experienced a $500,000 savings [22]. A study done in Bergen County, New Jersey, elicited similar findings: fuel oil savings were over $70,000 per year; savings on the purchase of passenger vehicles averaged $525 per car, or over $26,000 total; the savings to taxpayers was estimated at over $100,000 per year [23]. In addition to the incentive provided in the way of actual savings, the ABA Model Procurement Code suggests that any money received by one public purchasing agency for providing services, supplies, or facilities to another be recovered into the agency's budget and not turned over as general receipts of the public body [24]. Consequently, the financial benefits of cooperative purchasing agreements accrue directly to the responsible agency.

In addition to time and money savings, the implications of an intergovernmental purchasing strategy on energy conservation are several. First, allowing state governments to purchase items for smaller local and regional units should reduce unnecessary waste—the kind of waste produced when small governmental units, in order to benefit from volume discounts, are forced to purchase materials in such large quantities that the materials are either never used or are used without consideration of the need for conservation.

Second, allowing cooperative or joint use of supplies, equipment, and facilities may eliminate unnecessary duplicate purchasing of equipment and services that are infrequently or intermittently used. While emergency vehicles, such as fire-fighting equipment and ambulances, may not lend themselves to such sharings, equipment used to service and maintain a community on a predictable schedule, such as solid waste collection equipment, leaf vacuums, and road maintenance and repair equipment, could receive more efficient use if shared by neighboring municipalities. Of course, the financial benefits of joint purchase and use of this type of equipment may be partly offset by the need for more frequent repair or earlier replacement of the equipment, but the combined assets available to joint purchasers initially should make it possible to purchase more efficient and durable equipment [25]. As mentioned earlier, this type of joint use may create substantial demand for more types and varieties of energy-efficient equipment and thus serve as an incentive to manufacturers to develop new, more efficient equipment. Along the same lines, storage facilities could be jointly used.

Finally, by allowing governmental units to purchase supplies and equipment from each other, a cooperative purchasing policy provides a channel for disposing of surplus or reusable items and equipment [26].

The incentives of time, money, and energy savings built into any cooperative purchasing program should make it an attractive idea to any state or local government agency. Currently, the impediment to instituting such a policy is the lack of clear legislative or regulatory authority to engage in cooperative purchasing agreements. Statutory authorization is often needed [27]. First, there is a need for more comprehensive laws: all state and local purchasing laws should explicitly provide authority to purchase cooperatively with other jurisdictions [28]. The ABA Model Code Draft and the California and New Jersey statutes are good examples of comprehensive and effective legislation [29]. A second need is for local governments to overcome their fear of subordination to larger governments and their preference for local suppliers.

Several field studies of cooperative purchasing practices have identified the most successful methods of introducing such a program. In Detroit, researchers found that grouping local units was more readily accepted than a large unit organizing the cooperative purchasing program [30]. A study of the Washington, D.C., Maryland, and Virginia areas also supported the idea of starting with a relatively small group [31]. Both these studies recognized and affirmed the need for a centralized, specialized guiding force, though the Detroit study suggested that the practical effectiveness of the program could be demonstrated to local units by making a one time joint purchase of some commonly used item [32]. Similarly, the metropolitan Washington area study suggested limiting initial purchases to large quantity use commodities that do not present special difficulties [33]. A final implementation strategy concerns educating the participating or interested governmental units by arranging and sponsoring periodic meetings to exchange information and problems [34]. Use of these suggested implementation strategies should help encourage receptive attitudes toward intergovernmental cooperative purchasing. State and local governmental support and commitment to such a program should help maintain positive attitudes and cooperative practices.

NOTES TO CHAPTER 5

1. *See*, Council of State Governments, *State and Local Government Purchasing* (Lexington, Kentucky, 1975) p. 14.1.

2. *State and Local Government Purchasing, supra* note 1; American Bar Association, Coordinating Committee on a Model Procurement Code, *Model Procurement Code—Preliminary Working Paper No. 1* (hereinafter cited as *ABA Model Code—No. 1*), Section 8 (Washington, D.C.: ABA, 1976).

3. Robert M. Belmonte, "Another Look at Large Scale Intergovernmental Cooperative Purchasing," *Journal of Purchasing* 8, no. 1 (February 1972): 36.

4. Tex. Educ. Code § 21.161 (1949).

5. *State and Local Government Purchasing, supra* note 1 at 14.2.

6. Telephone interview, Ms. Patsy Anderson, Council of State Governments, Lexington, Kentucky, December 1, 1976. Figure is based on 1972 survey.

7. Belmonte, *supra* note 3 at 41.

8. *State and Local Government Purchasing, supra* note 1 at 14.2. *See also* George W. Jennings, *State Purchasing: The Essentials of Modern Service for Modern Government* (Lexington, Kentucky: Council of State Governments, 1969), p. 62.

9. Information from State Purchasing Agency, Department of General Services, State of Illinois.

10. *State and Local Government Purchasing, supra* note 1 at 14.2.

11. Cal. Gov't Code § 14814 (West Supp. 1976); *See also, Ibid.*

12. *State and Local Government Purchasing, supra* note 1 at 14.2.

13. Belmonte, *supra* note 3.

14. N.J. Stat. Ann. § 40A:11–15 (1975), *as amended*, N.J. Session Law Service, 196th Legislature, 2d sess., ch. 353, 1975 (April 1976).

15. *State and Local Government Purchasing, supra* note 1 at 14.4.

16. Raleigh F. Steinhauer, "Intergovernmental Cooperative Purchasing: The Wave of the Future?" *Journal of Purchasing* 8, no. 3 (August 1972).

17. Belmonte, *supra*, note 3 at 42.

18. *ABA Model Code—No. 1, supra* note 2, § 8–108.

19. *Ibid.*, § 8–109(3).

20. *Ibid.*, § 8–109(4).

21. Kansas Advisory Council on Intergovernmental Relations (KACIR), "A Study of Directors of State Purchasing," p. 3 of Question 2g; and Special MML Staff Report, "Cooperation in Public Purchasing," *Missouri Municipal Review* (December 1972), p. 7.

22. KACIR, *supra* note 21; and Special MML Staff Report, *supra* note 21.

23. Belmonte, *supra* note 3 at 43–46.

24. *ABA Model Code—No. 1, supra* note 2, § 8–106.

25. See generally Chapter 1 of this book.

26. See generally Chapter 8 of this book; and Norman L. Dean, *The Environmental Law Institute State and Local Energy Conservation Project, Energy Efficiency in Industry: A Guide to Legal Barriers and Opportunities* (Cambridge, Massachusetts: Ballinger, 1977).

27. *State and Local Government Purchasing, supra* note 1 at 14.1.

28. *Ibid.*, p. 14.3; *See also* Belmonte, *supra* note 3 at 36.

29. *See ABA Model Code—No. 1, supra* note 2: N.J. Stat. Ann. §40A:11—10 et seq., *as amended*, N.J. Session Law Service, 196th Legislature, 2d sess., ch. 353, 1975 (April 1976).

30. Clyde T. Hardwick, "Regional Purchasing: A Study in Governmental Cooperative Buying," *Journal of Purchasing* 5, no. 4 (November 1969): 13—19.

31. Steinhauer, *supra* note 16.

32. Hardwick, *supra* note 30 at 16.

33. Steinhauer, *supra* note 16.

34. *See* Hardwick, *supra* note 30; and *State and Local Government Purchasing, supra* note 1 at 14.4.

Chapter 6

Energy Audits

INTRODUCTION

This and subsequent chapters turn from the discussion of purchasing to an equally important consideration: What becomes of real and personal property held by governments, and How is energy conserved or wasted in the process? Few, if any, governments can afford to replace all their property at the same time. Technologically up-to-date products are acquired only as older products wear out or become obsolete. Further, even if a government could afford to replace all its property with "state of the art" energy-efficient items (the most advanced available), further technological advances would soon render this new property ineffi-cient—relative to the newly advanced state of the art. Should the government then "rereplace" all its property with yet more technologically advanced models? If the answer were yes, governments would likely establish agencies to retard technological advancement; such foot-dragging would become not only sanctioned, but also insti-tutionalized.

This dilemma introduces the premise underlying Chapters 6 and 7: very few items in government possession at a given time will be as energy-efficient as the most effective product available, and yet replacing all the "laggards" immediately will not be economically feasible. No industry, much less a government, can keep in perfect step with technology. Coping with this perpetual technological lag is the subject next treated.

Whether they are considering purchasing a particular item or discarding it, governments should peruse these options: (1) disposing of the item (in an ecologically sound manner—see Chapter 8, and acquiring one that is more energy conserving; (2) modifying the item to make it more energy-efficient; and (3) managing the item and the employees working with it so as to minimize its energy use. The third method is, of course, applicable in the first two situations as well, but is of particular importance when a government decides to retain an inefficient item.

Chapter 6 suggests a method, an energy audit, for deciding which option is appropriate in a particular instance. Only by keeping an "energy ledger" as a record of how it uses its energy can a government mount an effective conservation campaign. By monitoring the energy use of each major item of property (primarily buildings and motor vehicles), governments can determine which items are used most and which are functioning efficiently. In addition, auditing energy consumption over a period of time provides one measure for evaluating the success of the conservation program.

When its replacement is not justified, government property can often be modified to increase its energy efficiency. Modification, or retrofitting, the subject of Chapter 7, is a strategy of particular salience when it is applied to buildings. The former administrator of the United States General Services Administration was on target in pointing out that 85 percent of the office buildings that will be around in the year 2000 had already been built in 1975 [1]. Retrofitting will thus play a much more significant role than procurement in near term building energy conservation.

Finally, the latter part of Chapter 7 turns from the discussion of hardware to a brief consideration of the government employees who use government property. This discussion again shows how the wasteful use of energy brought about by human neglect can quickly nullify the benefit of mechanical energy efficiency. Key points of energy conservation management are cited, and managerial credibility is emphasized. A discussion of means to influence employees to use energy-efficient transportation for commuting is also included.

Governments, because of their size and heterogeneity, can in many contexts be thought of as microcosms of society. Because they employ their "constituents," governments have much greater control over the functioning of their internal "societies" than they have over society as a whole. As is suggested throughout this book, these two facts combine to create special potential for governments as agents of conservation. First, as major energy users *within* society, governments can achieve internal energy savings great enough to be felt on a societal level. Second, governments, as minisocieties, can demonstrate the feasibility of innovative energy conservation methods and can thus provide models for families as well as for private enterprise.

AUDITING ENERGY

An energy audit, in the context of this book, is a survey of the energy use of a government unit [2]. The audit should reveal the total amount of energy used by the unit over a given period of time and should also break down the total according to amounts used by individual energy-consuming items. Buildings and motor vehicles are prime targets for individual treatment. The more detailed the audit is, i.e., the more it breaks the total down into use by individual items, the more useful it will be in pinpointing areas of energy waste.

An energy audit is a highly appropriate first step in developing an energy conservation program. Audits suggest which items should be first in line for efficiency improvements. Where funds are short, accurate identification of priorities can be crucial to program success. And where the political acceptability of a conservation strategy hinges on rapid results, audits can help choose areas where implementation can bring the most dramatic savings.

Energy audits can also be used to measure the effectiveness of an energy conservation program. Results of the initial audit serve as a baseline of energy consumption from which improvement can be measured. Results may also be compared to energy budgets established by various conservation entities [3]. Subsequent audits, which should be performed frequently and at regular intervals, will measure the amount of energy being saved and will also point out areas where optimal savings are not being achieved.

A variety of auditing methods have been developed by public and private entities. Only the simplest method will be discussed here.

Buildings

As noted above, buildings owned by governments are prime targets for auditing. While development of an optimal conservation plan may require the use of computer analysis of consumption (see appendix), a preliminary evaluation of overall efficiency can be performed at nominal administrative cost. Preliminary analysis of a series of buildings can help determine which ones are most inefficient, and consequently possess the greatest potential for energy savings. These buildings, if sufficiently large, are also the most promising candidates for more complex analysis.

The preliminary audit consists of merely calculating the average number of BTU consumed annually by one square foot of floor space (BTU/ft^2/yr). The average includes energy for heating, cooling, ventilating, and lighting. This measure, which has been called both the energy utilization index [4] and the energy performance index [5], is a commonly used indicator of building energy efficiency. It is also used to prescribe efficiency standards for the design of new buildings [6].

The following information is required to determine the energy utilization index for a building:

1. the number of square feet of floor space,
2. the amount of fuel used, as shown on utility and other energy bills for a specific time period (several months to one year), and
3. the number of "degree days" for the same time period [7].

The number of "degree days" for any time period is a measure of weather conditions for that period [8]. It has been determined that energy consumption of a building's heating and cooling systems varies directly (if roughly) with the difference between the outdoor, mean daily temperature and 65°F [9]; the number of degree days shows that difference. Heating degree days and cooling degree days are calculated in the same fashion, although factors other than temperature, such as wind and sunlight, have a significant effect on the amount of energy needed for cooling [10]. If the maximum temperature for a day is 92°F and the minimum is 62°F, then the mean is 77°F, and there are twelve degree days (77° − 65°) for that day. When, over a period of a year, this calculation is performed for each day of a month, and the degree days within each month are totaled, the result is a useful comparison of the load on heating and cooling systems during different months. For example, a thirty day month consisting entirely of days having twelve degree days is a 360 degree day month.

The commonly used measure of building energy efficiency, as stated above, is the average BTU per square foot of floor space (per year or other time period), or energy utilization index. Given that weather conditions, and therefore heating and cooling load factors, fluctuate substantially from month to month and from year to year, accurate measurement of efficiency requires a means of adjusting the figures for these comparable time periods to account for the fluctuations. Otherwise, if energy use for a month with a high load factor were compared with that of one with a light load, it would create the false impression that the building's systems were functioning more efficiently during the light load month. Degree days may be used in two ways to provide this adjustment: (1) they may be presented independently of the energy utilization index, or (2) they may be used in calculation of the index. In the latter case, the index will be expressed in BTU per square foot per degree day. (Information about degree days for a given area can usually be obtained from the United States National Weather Service.)

Calculation of the index using degree days is as set out below; a sample worksheet for performing the calculation is provided in the appendix.

1. From the energy bills for a recent period, determine the amount of each type of fuel used for that period.
2. Convert the amounts of fuel used into BTU, as shown in the appendix [11].
3. Add the BTU from all fuel sources.
4. Divide the total BTU by square feet of floor space.
5. Divide the quotient by the number of degree days for the period.

The results of this simple calculation can be used to determine if further, more complex and costly analysis is justified in order to maximize the effect of the conservation strategy.

School officials may wish to avail themselves of a computerized analysis of the energy efficiency of the buildings under their control. The program was developed under a project funded by the Federal Energy Administration. In return for filling out a form describing the physical characteristics and energy use of their buildings, schools can receive a printout evaluating their energy use as compared to an energy-efficient model and suggesting the amount of energy that they might be capable of saving [12]. Interested officials should contact John Boice, Educational Facilities Laboratories, Inc., 3000 Sand Hill Road, Menlo Park, CA 94025.

The Energy Conservation and Production Act

The recently enacted Energy Conservation and Production Act (ECPA) [13] authorizes federal funding of "supplemental State energy conservation plans" which must include "procedures for encouraging and for carrying out energy audits with respect to buildings and industrial plants within such State" [14]. These plans, as set out in ECPA, constitute an amendment to the sections of the Energy Policy and Conservation Act that provide for the original state energy conservation plans [15]. "Energy audits," as defined in ECPA, actually go several steps beyond the audit process described in this chapter: "(3) The term "energy audit" means any process which identifies and specifies the energy and cost savings which are likely to be realized through the purchase and installation of particular energy conservation measures or renewable-resource energy measures . . ." [16]. "Energy audits" under the ECPA extend beyond the audit process and into what might be called a preretrofit analysis. This type of analysis is discussed at the beginning of the next chapter.

As this book goes to press, it is still somewhat difficult to judge the import of the recent act, but states are urged to follow the development of guidelines for implementation of the supplemental conservation plans. These guidelines are being developed by the Federal Energy Administration and it may be that some or all of the cost of developing and administering energy audits of government buildings and certain other buildings in the state can be paid by the federal government [17].

Motor Vehicles

Motor vehicles constitute the second major category of candidates for energy audits. The method is fairly obvious. Operators of government vehicles should be required to submit monthly reports of miles traveled, both in traffic and on the open road, and of the amount of gasoline used. If vehicles are not assigned to individuals, a log should be kept for the vehicle, and each driver should be required to record gasoline use and miles traveled.

The information gained from the energy audit may suggest the need for more efficient vehicles or for more frequent and comprehensive maintenance. It might also be used in programs to encourage government employees to strive for more efficient operation of their own automobiles [18].

NOTES TO CHAPTER 6

1. United States General Services Administration, *Energy Conservation Guidelines for Existing Office Buildings* (Washington, D.C.: GSA, 1975), Introductory letter by Arthur F. Sampson, former administrator, GSA.

2. *See* Public Technology, Inc. (PTI), *Energy Conservation: A Technical Guide for State and Local Governments* (Washington, D.C., March 1975), pp. 7–11 and 41–42.

3. For example, the United States General Services Administration (GSA) has set an energy budget of 75,000 BTU/Ft^2/yr for retrofitted buildings. GSA, *Energy Conservation Guidelines for Existing Office Buildings* (Washington, D.C., 1975), p. 1–5.

4. National Electrical Manufacturers Association (NEMA), *Total Energy Management: A Practical Handbook on Energy Conservation and Management* (Washington, D.C., December 1975), p. 7–9.

5. *Florida Energy Conservation Manual,* prepared for state of Florida, department of general services (Tampa, Florida, May 1975), p. 15.

6. *Ibid.*; U.S. General Services Administration *Energy Conservation Design Guidelines for New Office Buildings,* 2nd ed. (Washington, D.C., July 1975), pp. 1–1 – 1–2.

7. PTI, *supra* note 1 at 9.

8. *See* NEMA, *supra* note 2 at 8.

9. *Ibid.*

10. *Ibid.*

11. Note that a better comparison among fuels is obtained if the figure used in the conversion represents the amount of fuel expended at the power plant rather than at the building. For example, 3 BTU are required to deliver 1 BTU of electricity to the building.

12. Conversation with Mr. Jean Miller of the Federal Energy Administration, November 18, 1976; and "Conservation: School Buildings are Target of Government Energy-Saving Experimental Programs," *Energy Users Report,* no. 159 (Washington, D.C.: Bureau of National Affairs, August 26, 1976), p. E–3.

13. Energy Conservation and Production Act (ECPA), §§431–32, Pub. L. No. 94–385, 90 Stat. 1125 et seq. (Aug. 14, 1976), *amending* Part C of title 3 of the Energy Policy and Conservation Act, §§361–66, Pub. L. No. 94–163, 42 U.S.C. §§6321–26 (Dec. 22, 1975).

14. 42 U.S.C. §6327.

15. 42 U.S.C. §§6321–26, *as amended by* ECPA, §§431–32.

16. 42 U.S.C. §6326(3).

17. The first Federal Energy Administration document issued with regard to the Energy Conservation and Production Act appeared in the Federal Register on November 1, 1976; the document refers to "energy conservation measures" and "renewable-resource energy measures." 41 Fed. Reg. 47998–48000 (Nov. 1, 1976).

18. Employee programs are discussed in Chapter 7 of this book.

Appendix to Chapter 6

AN ENERGY AUDIT PROCEDURE
DERIVED IN PART FROM FEDERAL ENERGY ADMINISTRATION
(FEA), OFFICE OF ENERGY CONSERVATION AND ENVIRONMENT,
GUIDELINES FOR SAVING ENERGY IN EXISTING BUILDINGS:
BUILDING OWNERS AND OPERATORS MANUAL, ECM 1,
CONSERVATION PAPER NUMBER 20 (WASHINGTON, D.C.:
FEA, JUNE 1976), pp. 82–88.

The following steps are to be used in the preparation of a profile of existing energy consumption for a particular facility. First, the gross number of BTUs of energy use per square foot of floor area is determined, and then this number is broken down into usage by various subsystems, such as heating, hot water, etc. Where possible, auditors should break down usage by individual machines, since energy conservation measures will be considered by reference to replacing or modifying these units, but this degree of breakdown is only suggested here. The overall usage figure and the breakdowns will provide "baselines" of use against which to measure energy savings and program success.

Step 1: Determine Gross Square Feet of Floor Area

Filling in the blanks under Figure A6–1, below, provides the gross floor area of the building under audit. The validity of using this figure assumes fairly homogeneous and consistent energy usage throughout the building. If there are rooms or areas within the building that use little or no energy (such as unheated storage rooms), these may be subtracted from the gross figure—provided such areas are adequately insulated from other space-conditioned areas.

Figure A6–1.   Gross Square Feet of Floor Area

---

1. Length of Building   _____ ft
2. Width of Building   _____ ft
3. Number of Floors   _____
4. Floor Area, gross ft^2 (line 1 × line 2 × line 3) _____ ft^2

**Step 2: Determine Gross Number of BTUs Consumed
per Year per Square Foot of Gross Floor Area**

Gather from your monthly utility and fuel suppliers' bills the annual usage of
energy in gallons of oil, cubic feet of gas, pounds of propane, tons of coal and
kilowatt hours of electricity. Record the gross yearly quantity of fuel and power
in Figure A6–2, Column A, below. Convert the units of fuel and electricity to
equivalent BTUs by multiplying the quantities of fuel and power by the appropriate conversion factor in Column B, and list the result in Column C. Total the
gross number of BTUs and record the result in line 7, below. Divide this number
by the gross square feet of floor area (as determined in Figure A6–1 and recorded on line 4 of Figure A6–1), and enter the result on line 8. This is the primary piece of information about your building's energy usage, and can be used
in establishing an energy budget for the building. A realistic energy budget goal
for existing buildings may be as low as 75,000 BTUs/ft^2/yr for office buildings
and stores, 60,000 BTUs/ft^2/yr for schools, and 35,000 BTUs/ft^2/yr for reli-
gious buildings.

Figure A6–2.   Gross BTUs/ft^2/Year

	_A_	_B_	_C_
		Conversion Factor	Thousands of BTUs/yr.
	_____	× 138 (*) =	_____
1. Oil—gallons		× 146 (**) =	_____
	_____	× 1.0 (†) =	_____
2. Gas—cubic feet	_____	× 0.8 (††) =	_____
3. Coal—short tons	_____	× 26000 =	_____
4. Steam—pounds × 10^3	_____	× 900 =	_____
5. Propane Gas—lbs	_____	× 21.5 =	_____
6. Electricity—Kw hrs	_____	× 3.413 =	_____
7. Total BTUs × 10^3/yr	. . . .		_____
8. BTUs × 10^3/Yr Per Square Foot of Floor Area . . . . _____			
(line 7 ÷ line 4, Figure A6–1)			

---

*Use for No. 2 oil; **Use for No. 6 oil; †Use for natural gas; ††Use for manufactured gas.

**Step 3: Determine BTU Consumption per Year
per Square Foot of Gross Floor Area for Systems —
Heating, Hot Water, Cooling, and Lighting**

Next break down the average annual BTU consumption by system, using Figures A6–3 through A6–5, below. The duration and severity of winter varies from year to year, as this past year (1977) has demonstrated dramatically, and affects the quantity of heat required. To obtain the most accurate results, five or more years of energy consumption should be averaged to determine fuel consumption for a typical year for heating. If records are only available for the previous year's fuel consumption, these can be corrected to a typical year by dividing the actual fuel consumption by heating degree days actually experienced for that year, then multiplying by the *average* yearly degree days for the building's geographic location, as discussed in Chapter 6. The heating degree days for the year corresponding to actual fuel consumption can be obtained from the weather bureau, or from a local fuel supplier.

Fuel bills often do not differentiate between the end use for heating or other purposes, and an adjustment must be made. If oil, gas, or coal is the primary fuel, and is used for both heating and domestic hot water, the usage should be broken down between the two. The space heating load occurs in the winter, but the domestic hot water load is continuous for the whole year at a rate that can be assumed to be constant. To determine the amount of the monthly fuel bills that can be attributed only to heating, select one average winter month's consumption, subtract one average summer month's consumption and multiply the answer by the total number of heating months. The difference between total fuel used and heating fuel use will be the domestic hot water energy consumption.

If the building is heated by electricity and the total electrical usage of the building is metered and billed in a lump sum, the bill will include energy for heating, lighting, and power. To arrive at the amount of electricity used for heating only, it is necessary to assess the quantity used for lighting and power and subtract this from the total billing. In small buildings, a quick assessment of the electricity usage for lighting can be made by counting the number of lighting fixtures and multiplying the wattage of each lamp and the average number of hours that these are switched on during the heating season. This will give the total number of watt-hours consumption that can be attributed to lighting. Divide watt-hrs. by 1000 to get kilowatt-hrs. Similarly, a survey can be made of all electrical motors that are in use during the heating season and their nominal horsepower rating (multiplied by .800) to determine the approximate amount of electricity in KWH used for each hour of running. (This formula assumes an efficiency of 93% for electric motors). The KWH should then be multiplied by the number of hours of operation during the heating season to determine the total kilowatt hours that can be attributed to power. The sum of the kilowatt hours assessed for lighting and power should then be subtracted from the total power consumed by the building for the heating season, to determine the amount used for heating. In large, complex buildings where simultaneous heating and cooling are likely to occur, you should seek professional help to prepare a more accurate analysis of energy flow, if your maintenance staff is unable to do so. To determine the energy used for lighting and for power for the entire year, the same method of determining energy use in the heating season, described above, can be used for a 12-month period.

To determine the amount of energy used for air conditioning, estimate the energy for fans and pumps as outlined above. For electric driven refrigeration units the KWH can be estimated by deducting the energy used for lighting and other motors from the June, July, August, and September electric utility bills.

Figure A6–3. Annual Fuel and Energy Consumption for Heating

	A	*B* Conversion Factor	*C* Thousands of BTUs/yr.
		× 138 (*) =	
1. Oil—gallons		× 146 (**) =	
		× 1.0 (†) =	
2. Gas—cubic feet		× 0.8 (††) =	
3. Coal—short tons		× 26000 =	
4. Steam—pounds × 10^3		× 900 =	
5. Propane Gas—lbs		× 21.5 =	
6. Electricity—Kw hrs		× 3.413 =	
7. Total BTUs ×	.... =		
8. BTUs × 10^3/Yr Per Square Foot of Floor Area .... (Line 7 ÷ line 4, Figure A6–1)			

*Use for No. 2 oil
**Use for No. 6 oil
†Use for natural gas
††Use for manufactured gas

Figure A6—4. Annual Fuel and Energy Consumption for Domestic Hot Water

	A	B Conversion Factor		C Thousands of BTUs/Yr.
1. Oil—gallons		× 138 (*)	=	
		× 146 (**)	=	
2. Gas—cubic feet		× 1.0 (†)	=	
		× 0.8 (††)	=	
3. Coal—short tons		× 26000	=	
4. Steam—pounds × 10^3		× 900	=	
5. Propane Gas—lbs		× 21.5	=	
6. Electricity—Kw hrs		× 3.413	=	
7. Total BTUs/Yr × 10^3	. . . .			
8. BTUs × 10^3/Yr Per Square Foot of Floor Area . . . . . . (line 7 ÷ line 4, Figure A6—1)				

*Use for No. 2 oil
**Use for No. 6 oil
†Use for natural gas
††Use for manufactured gas

Figure A6–5.   Annual Fuel and/or Energy Consumption for Cooling
(Compressors and Chillers)

	<u>A</u>	<u>B</u> Conversion Factor	<u>C</u> Thousands of BTUs/Yr.
a. *if absorption cooling*		× 138 (*) =	
1. Oil—gallons	_____	× 146 (**) =	_____
		× 1.0 (†) =	_____
2. Gas—cubic feet	_____	× 0.8 (††) =	_____
3. Coal—short tons	_____	× 26000 =	_____
4. Steam—pounds × 10^3	_____	× 900 =	_____
5. Propane Gas—lbs	_____	× 21.5 =	_____
6. Total BTUs/yr × 10^3	. . . .		_____
7. BTUs × 10^3/Yr Per Square Foot of Floor Area . . . . (line 6 ÷ line 4, Figure A6–1)			_____
b. *if electric cooling*			
8. Electricity—KWH	_____	× 3.413 =	_____
9. BTUs × 10^3/Yr Per Square Foot of Floor Area . . . . (line 8 ÷ line 4, Figure A6–1)			_____

*Use for No. 2 oil
**Use for No. 6 oil
†Use for natural gas
††Use for manufactured gas

※ *Chapter 7*

Modification, Management, and Maintenance—The Ownership Strategies

MODIFICATION (RETROFITTING)

Where an energy-wasteful item owned by a government has a remaining useful life of significant duration, it should be modified to improve its efficiency. This approach is often called retrofitting. Although the practice is applicable to any energy-consuming item in government possession, discussion focuses on the modification of buildings. In some ways, retrofitting is a purchasing strategy, since the parts and labor must be procured in much the same way as the original item was obtained. Retrofitting is discussed here because this chapter's focus is upon items *already* in government possession.

For purposes of energy conservation action, government possessions may be categorized into (1) those to be replaced by purchases of more efficient items, (2) those to be kept as is, and (3) those to be kept and modified to improve efficiency [1]. The first step in a retrofitting program is discovering what government items fall into the third category, which goes back to the concept of energy audits examined in Chapter 6.

With well-defined retrofit methods it is possible to carry analysis much further than that suggested in Chapter 6. Life cycle cost analysis, applied in Chapter 1 to purchasing, can also be used with respect to property management to identify modifications that will yield the highest returns [2]. Returns can be measured either in terms of the annual BTU savings per dollar spent on modification or, by translating the BTU savings into dollars, in terms of dollars saved in the future per dollar expended now.

Factors Influencing the Decision to Retrofit

It should be stressed that the decision to retrofit a particular item should not be isolated from other energy conservation decisions. As indicated above, the preliminary analysis of the energy efficiency of government property could suggest one of three possibilities: retrofitting, replacement, or retaining the item for limited use. Further, there are many possible means for achieving the goal of conservation; only by systematic analysis of all available strategies and of their applications to specific situations can the optimal mix be attained. This caveat is included because the discussion below concentrates on retrofitting alone and does not consider other possibilities or combinations.

While retrofitting is a short term strategy in relation to a specific item, it is by *no means* a *temporary* strategy. Medical first aid might also be considered a short term strategy, because it is applied immediately and further attention is usually required, but it cannot be regarded as a temporary strategy because new patients will always be in abundance. Similarly, retrofitting is a strategy to help the government "make do" with an inefficient item until its replacement can be justified; however, to the extent that technology speeds obsolescence or that increased energy costs justify greater expenditures for efficiency, government property will always be *relatively* inefficient. The decisionmaking process should therefore be institutionalized, so that some one entity is continually responsible for mating the best retrofit candidates with the most remunerative retrofit methods.

Ideally, the process should include the following components:

1. A continuing energy audit of each major energy-consuming item in government possession. The audit should include both the energy efficiency and the usage level of the item.
2. A continuously updated "catalog" of retrofit alternatives for each type of item (buildings, etc.), with a means of estimating cost and potential energy savings in various applications.
3. A comparison of the ratio of the energy savings during the remaining life of the item to the cost of application for the range of alternatives applicable to that item (essentially, a life cycle costing for each possible combination).
4. Finally, adoption of the modifications that are shown to have the highest ratio of energy savings to cost. The number of items retrofitted at any one time will in part be determined by the current availability of funds for this purpose and by the government's assessment of how much energy conservation is worth [3]. Where the ratio of savings to costs is high, governments empowered to

do so should consider borrowing funds needed to retrofit, especially where savings will be realized in the near term.

As suggested in point 3, this process results in a slightly complicated form of life cycle costing. As compared to the life cycle costing discussed in Chapter 1, the process here described is more complex because of the need to project the effect of the modification upon the efficiency of an existing item. Some analysis will often be time-consuming and expensive. Some combinations will probably be quickly ruled out by simple preliminary analysis. For the rest, the savings may be well worth the cost. Ohio State University reports that the savings from modifying only six of its buildings will cover the modification costs in eight months, and then produce annual savings of over $300,000 [4].

Possible Retrofit Methods
The rest of this chapter will briefly introduce the reader to some of the technical modifications applicable to buildings.

The Building Envelope: Roof, Walls, and Floor [5]. Heating and cooling of buildings accounts for 21 percent of U.S. energy consumption. A great deal of this energy is wasted because of transmission of heat through the building envelope. The building envelope consists of windows, doors, the roof, walls, and the floor—i.e., any barrier having the building's interior space on one side and the outdoor environment on the other. When the building is being heated, energy waste is due to transmission of heat to the outside; when it is being cooled, waste is due to transmission of heat to the inside. Reducing the rate of transmission will achieve substantial energy savings.

Heat transmission is primarily affected by the resistance of the various components of the envelope to heat flow. The resistance varies with both the nature of the component (wood, steel, asbestos, etc.) and its thickness. One way of expressing this quality is in terms of "U" values or "R" values, which are reciprocal. U values express the capacity of a material for transmitting heat; R values express its resistance to heat transmission. The U value of a particular building component denotes the number of BTU that will pass through one square foot of that component in one hour for each degree Fahrenheit of difference in temperature between either side of the component.

The U value thus provides a conceptual handle for the problem of reducing heat transmission through the building envelope. Reducing the U value carries with it a directly proportional reduction in heat transmission and can save bundles of energy.

To reduce the U value, insulate the building, i.e., make the envelope thicker and of materials with a high resistance to heat transmission. Obviously, this is more easily done (and the costs more easily justified) during construction than after a building is completed. Many existing buildings, however, have such poor insulation that it is not difficult to find opportunities for cost-justified retrofitting.

A recent study for the Federal Energy Administration offers three alternatives for increasing wall insulation:

1. Add insulation to outside walls. This method is easiest for one and two story buildings and can be done without disturbing occupants.
2. Add insulation to inside walls. Multistory buildings do not present an impediment to this method, but the method will disturb occupants and may be hard to integrate with existing wall fixtures such as receptacles and ducts.
3. Fill the space between inside and outside walls, where such a space exists, with granular insulation or polyurethane. This method requires drilling holes through the wall, but will not alter the building's appearance or greatly inconvenience the building's occupants [6].

Of these methods, filling the wall cavity with polyurethane is the cheapest alternative, while adding rigid insulation to outside walls is the most expensive [7].

Roofs present frequent opportunities for retrofitting, since they are repaired more often than any other building component. Regular roof repair can be accompanied by the spraying on of foam or the installation of rigid insulation, and economies will be realized by combining these tasks into a single operation [8]. Alternatively, where repair is not imminent, rigid insulation or spray on foam may be added to the underside of the roof if that area is accessible.

Heat transmission through floors varies according to whether the floor is suspended or "slab on grade" (in direct contact with the earth) [9]. Where a floor is suspended, it transmits heat to or from the space below in much the same way as a wall [10]. While it may be impracticable to add insulation to the top of a floor, spray or rigid insulation can be applied to the underside. Thanks to the insulating effect of the earth, slab on grade floors lose very little heat through centers and require insulation only around their perimeters [11]. Such insulation should be positioned vertically on the outside edge of the floor.

In addition to the capacity to transmit heat, there are two characteristics of building envelope components that substantially affect the energy efficiency of the building: (1) their thermal inertia, and (2) their tendency to absorb solar radiation.

Building components of large mass have the capability of storing heat and thus moderating the effect of rapid temperature changes outside the building [12]. This quality is called thermal inertia. Even though both have the same U value, a low mass wall (ten to twenty pounds per square foot) will have a 2 percent greater heat loss than a high mass wall (eighty to ninety pounds per square foot) [13]. High thermal inertia is more important in regions with extreme and rapid daily temperature variations than in regions having a moderate and more stable climate [14].

Walls and roofs absorb solar radiation in relation to the colors of their exterior surfaces. The darker colored surfaces, said to have high absorption coefficients, absorb more solar energy than do the lighter ones, which have low absorption coefficients [15]. Although it is very easy to change a building's color, deciding whether a color change is desirable is a more difficult problem than predicting the value of insulation. Insulation is of universal benefit, since it increases the efficiency of both heating and cooling. Increasing a building's absorption coefficient, however, may save little heating energy and waste a lot of cooling energy. Decisions affecting a building's absorption coefficient should include consideration of climate, weather, and building orientation.

The Building Envelope: Windows. Except to the extent that daylight can replace artificial lighting, windows are likely to waste a great deal of energy. While there does exist an outdoor cafe in Washington, D.C., which is heated during the winter months, few architects would recommend that office space in temperate climates consist of open-sided platforms with great space conditioners capable of maintaining a constant temperature within the "office." If such an arrangement is seen as flagrant energy waste, then so should the construction of buildings with solid walls of windows. While the U values of opaque walls can be reduced to 0.04 or less, a single glass has a U value of about 1.13 [16]. Alternately stated, this means that the solid glass outside walls seen on many modern office buildings permit heat transfer, and thus waste energy, at over twenty-eight times the rate of a well-insulated opaque wall.

While it is the author's personal view that requiring people to work in windowless offices is, at best, a barbaric practice, it is obvious that improper fenestration is a major energy waster [17]. Approaches to reducing heat transfer through the windows of existing buildings include (1) blocking off part of the window area with opaque insulated barriers, and (2) adding second and even third thicknesses of glass.

Where existing windows are clearly in excess of that required for natural lighting of the interior, opaque thermal barriers may be permanently installed. In other cases, barriers may be designed for removal during hours of occupancy. Table 7–1 [18] suggests how barriers and multiple glazing can modify U values.

Additional layers of glass can markedly reduce heat transfer through windows, although even triple glazing permits about ten times the heat transfer of a well-insulated opaque wall. According to the United States General Services Administration, a single glass has a U value of about 1.13, double glass of 0.58 to 0.69, and triple glass of 0.36 to 0.47 [19]. Thus, while triple glazing is nowhere near as resistant to heat transfer as an opaque wall, three panes are about three times as effective as one pane. The cost of converting to even double glazing, however, can be very high. Each government must obtain its own quotations from local contractors, but the Federal Energy Administration estimates costs at $7.50 per square foot for glass alone and $14.00 per square foot for replacement of existing single panes with sealed double glazed units [20]. Even this high cost may be justified, however, particularly for inefficient buildings in high degree day regions.

In moderate climates, and particularly in those that require air conditioning for a significant part of the year, windows may present an additional threat to energy conservation by allowing sunlight to enter and trapping its heat inside. One solution to this problem is the installation of reflective glass on the sides of the building that have the most solar exposure. An effective alternative is the application of reflective film to existing glass [21].

Building Energy Load: Computerized Management. By installing computerized load management systems it is possible to reduce signif-

Table 7–1. **Effects of Insulation on U Values**

Insulation Thickness of Barrier (inches)	Composite U Value	
	(existing windows plus thermal barrier)	
	Single Glazing	Double Glazing
0.5	0.28	0.23
1.0	0.18	0.16
1.5	0.13	0.12
2.0	0.11	0.10

icantly the energy consumption and the utility bills of existing buildings and other major facilities. These computer systems are attached to all energy-consuming parts of a major facility (fans, lights, heating, and cooling, etc.) and are capable of shutting off various machines or combinations of machines for brief intervals. The brief shutdowns (1) reduce overall building energy consumption without noticeably altering the functioning of the building's energy systems and (2) reduce the peak load of the building for each billing cycle, thus reducing the utility bill.

The peak load is the highest level of kwh demanded by a facility at any one time during a given period. Even though this peak may be reached for only one second during a particular billing cycle, the utility bill is in part related to the peak. This results from the fact that electricity cannot feasibly be stored and consequently must be generated at the same time it is used; therefore, the higher the peak, the more generating equipment the utility company must normally hold in reserve to meet peak demand. This reserve generating equipment, which may be needed for only one second a year, costs a lot to purchase and maintain. Accordingly, it is reasonable to charge more for a billing period that contains a disproportionately high peak demand than for periods in which demand is at a nearly constant level. Those consumers whose usage patterns force the utilities to maintain the costly reserve capacity are thereby charged for the extra expense attributable to this cost [22].

Load management systems reduce the peak load by anticipating it and then shutting down nonessential equipment until the peak passes. This can save a great deal of money. Savings in electric bills have been estimated at between 10 and 25 percent for major facilities that have installed such systems [23]. A government contemplating use of computerized load management systems can engage an appropriate consulting firm to determine whether the savings generated at a particular facility will be large enough to justify the installation and maintenance costs of the system [24].

It is hoped that this brief discussion will stimulate further interest in retrofitting for energy conservation. This chapter, however, is not even an introductory primer. There are dozens of other methods, including installation of such things as heat pumps, shutters, and blinds; planting of trees; constructing vestibules at main entrances; and so on.

MANAGEMENT AND MAINTENANCE

The discussion thus far has equated energy conservation efforts at government facilities with the acquisition of more energy-efficient hardware. While adequate equipment is vital to economic use of energy, the most efficient machine in the world will waste energy if it is used needlessly or if poor maintenance is allowed to eat away at its efficiency. This part focuses on one of the most important elements involved in conservation, the manner in which human beings maintain and operate government property.

The discussion primarily concerns management relations and the problems of influencing employees to conserve energy; in addition, it considers the extent to which governments can encourage employees to save energy off the job, namely, in commuting to and from work. Also included is a list of published sources of energy conservation methods that can be implemented via the employment relationship. These are both general and specialized works; the methods recommended relate to maintenance as well as to reduction of usage.

While a treatise many times the length of this work could be written on the scope of the employment relationship, it seems safe to say that employers are empowered to require that employees who use energy-consuming government property use it (1) in an efficient manner and (2) as little as possible. This strategy area is at the same time the easiest and the most difficult one discussed in this book. The moment a government official decides that it is important to save energy, walks to a supervisor's office, and says, "Be sure everyone shuts off their lights before leaving tonight," the strategy is underway. But maintaining employee participation and cooperation at a high level, particularly where conservation methods create inconvenience and discomfort, can be quite difficult. Furthermore, participation and cooperation must be obtained without detracting from the primary mission of the department and without diminishing employee morale. Such a feat calls for sensitive, intelligent, and innovative management.

. This book will not deal with the intricacies of management expertise; it is assumed that the government agencies that might apply the strategies outlined here will already have well-developed management structures and methods that can be adapted to energy conservation management. Key features of a successful energy management plan might include: (1) high level management commitment to saving energy; (2) programs to instill a sense of individual participation and responsibility in all employees; (3) information feedback on success or failure of individual roles in the conservation program to each office and, if possible, to each employee; and (4) incentives to energy saving—carrots to go along with the sticks. Government employers should also check whether proposed plans comply with any applicable safety and health regulations [25].

There exists an historic rift between employers and employees, which intensifies with proximity to the top and base of the managerial pyramid. Some of this gap must be bridged in order to implement an effective energy program. Obviously, if employees conserve only when confronted with enforcement, or only "when the boss is looking," the program's effectiveness will be less than optimal. Managing by example is one way to close the rift; hypocritical pontificating is a sure way to widen it. It is critical to separate the status and convenience that may justifiably import to the job performance of a top official from the perquisites that are obvious symbols of energy waste. A department head's memorandum urging employees to use public transportation is unlikely to be well received by those who view his daily descent from a chauffered limousine. The perspiring employee who is summoned from his own 80° office to the boss's 65° chamber will scarcely be impressed with management's commitment to conservation.

On the Job Energy Management
and Conservation

In April and May of 1975, as a follow-up to the first two years of the Federal Energy Management Program, teams from the Federal Energy Administration and the U.S. General Services Administration visited federal installations nationwide. The report of these visits is contained in *Energy Conservation Site Visit Report: Toward More Effective Energy Management*; [26] it is an instructive tool for energy management at any level and is one of the few publications that discusses administrative problems of strategy implementation. There follows a summary of the report's findings and recommendations regarding implementation.

Top Management Commitment. "Human nature is such that most employee time and effort is directed toward those aspects of the job that are closely reviewed and about which management is concerned." This observation highlights the necessity for interest and commitment at high levels of the management organization. Top management who feel that the importance of their primary mission overrides that of conservation should remember that depletion of energy supplies could bring shutdown or sharp curtailment of their operations.

Line Management Accountability. The site visit teams found a high positive correlation between the degree of conservation responsibility delegated to midlevel management and actual reduction in energy use. Giving full responsibility to maintenance staff brought limited results. "The maintenance staff is often thought of as having little status in the organization, with no authority whatsoever over those who perform operational duties, even though the operations may use significant quantities of energy." The report suggests that providing *incentives* to management charged with conservation responsibility may be more productive than enforcement procedures.

Formal Planning. The superiority of performance orientation over prescriptive (design) orientation is apparent in areas other than purchasing (Chapter 1). The report notes that the General Services Administration's directive that initiated the Federal Energy Management Program included some prescriptive guidelines (e.g., it suggested thermostat settings for summer and winter). Many federal installations responded to the directives in letter only; they took only the steps that were spelled out. "More performance-oriented guidelines should be considered so that conservation coordinators can assess the actions necessary to achieve maximum reasonable savings." It is possible that programs will be more effective if the responsibility for *planning* as well as for execution is delegated as much as possible.

Monitoring. Monitoring the energy use of each office or department is a critical tool for effective management (see also Chapter 6). Feedback ensures compliance, provides incentive, and helps in the discovery of new areas of potential energy savings.

Use of Technical Expertise. An inventory of skilled personnel should be taken early in the development of an energy management program. In many cases, employees will have skills that can contribute to the program. Not only is this approach economical, but it provides opportunity for additional employee participation.

Employee Awareness. Achieving awareness in the work force requires a combination of instilling habits and overcoming resistance. It is a slow process, demanding consistent attention. The report suggests that management solicit energy conservation ideas from employees. "Energy conservation coordinators should be receptive to even the most improbable suggestions because a receptive attitude encourages people to involve themselves in working toward the solution to the problem."

Resource Support. Purchasing of new efficient equipment and retrofitting of existing units are topics dealt with already. The report states that in some areas of conservation, management strategies are nearing their potential for maximum effectiveness, and the need for retrofitting and energy-efficient purchases is growing urgent. This is particularly true for buildings and motor vehicles.

During 1975, representatives of the United States General Accounting Office (GAO), an investigating arm of the Congress, visited seventy-seven federal installations and monitored their energy management programs [27]. GAO reported that while facility management officials "had been active in attempting to conserve energy," much more could be done. Criticisms of program management at certain facilities included:

1. the total lack of a formal conservation program;
2. the failure to assign program management responsibility to a single individual or group, or the failure of an appointed individual to devote significant time to conservation efforts;
3. the lack of monitoring to the extent that the success or failure of the program was not measurable;
4. the lack of independent review of existing programs;
5. infrequent or incomplete inspections of temperature and lighting levels;
6. inadequate efforts to spur employee cooperation in conservation measures [28].

In general, the GAO report called for greater leadership and more aggressive promotion of energy conservation.

Encouraging Employees to Save Energy Off the Job

Off the job conservation by government employees was mentioned in Chapter 4 in the context of transportation and the siting of public buildings. Here the energy use area is again transportation, but the nexus between employees' commuting and their employment is *parking* [29]. By restricting available parking space for cars, a government can force its employees to use alternative, energy-efficient transport or to use automobiles in a more efficient manner. An ancillary strategy discussed here is government provision of vans for vanpooling.

Government employees have traditionally been provided with parking spaces, usually underpriced and often free of charge, and in most areas this practice continues [30]. To the extent that government-subsidized parking encourages employees who could use public transit to drive, or causes employees who could carpool to ride alone, it fosters substantial energy waste [31]. While recognizing that employees from isolated areas may have no transportation alternative to single occupant driving, the agency in charge of parking spaces should, in some rational way, regulate their use to promote the saving of energy.

A number of combinations of methods and goals are possible. If some existing spaces were blocked off and used for bicycle racks or green space, the reduction below the daily space usage level would force some employees either to find alternative parking or to use public transit. Mere reduction of spaces with no compensatory action, however, would only create confusion, inconvenience, and employee hostility; and those who succeed in the intense competition for berths may also be those who could just as easily use public transit. In addition, this method alone would fail to encourage carpooling, for while everyone's chances of getting a space would improve if everyone carpooled, a few individuals acting in isolation would not significantly enhance their chances by carpooling.

Another method, more palatable and therefore more productive, would be to charge enough for parking so that some employees will switch to public transit or join carpools in order to save money. This would avoid confusion and would allow employees freedom of choice of sorts, but would unduly penalize those with no access to public transit. However, special low rates could be provided for the handicapped and others with proven inability to use public transit or carpools. Another variation would be to set rates in inverse proportion to the number of persons in each vehicle. Existing spaces could then be phased out as this method brought about a reduction in demand.

A third method is to assign parking spaces billed at a monthly rate, and give preference to carpools on the basis of the number of members, i.e., a four member pool would have preference over a three member pool, etc. Again, limiting exceptions might be granted, as above. This method has several disadvantages. First, a monthly rather than a daily billing would encourage drivers to drive every day to get their money's worth, whereas they might otherwise be willing to take public transit occasionally. Second, once a space is assigned, will the carpool membership decrease? Although tallying ridership on a daily basis would be cumbersome, and made more complicated by sick or otherwise legitimately absent riders, some form of monitoring appears to be necessary since abuses of carpool incentive programs are growing notorious. In one program where parking preferences were obtained by submitting an application with signatures of other employee-riders, one popular employee was found signed up to ride to work in eleven different cars, with eight different origins!

The FEA site visit report discussed earlier notes that carpooling, a near necessity during the oil embargo, has fallen off markedly at government facilities [32]. The report cites two major disincentives to carpooling: (1) varied employee work schedules and (2) continual turnover, mainly at military installations [33]. In partial response, the report recommends that employees be permitted or even encouraged to adjust their working schedules to facilitate the formation of carpools. It also suggests that government facilities try to coordinate pooling with private companies located nearby.

Employee education programs are another factor essential to the implementation and maintenance of altered transportation patterns. Such programs should not fail to promote use of bicycles where that alternative is a practical possibility. The provision of secure bicycle storage would encourage employees to bike to work, and a bicycle plan might be made part of a physical fitness program. At large facilities, the installation of bicycle paths would enhance the convenience and safety of people-powered transit. Where possible, these paths should be integrated with regional or municipal bike lanes.

Obviously, many other variations of these general methods are possible. A detailed discussion of parking strategies is contained in Durwood J. Zaelke's *Urban Transportation and Energy Conservation.* Also, the U.S. General Services Administration has issued regulations for carpool parking at federal agencies, excerpts from which are included in the appendix [34].

Vanpooling is likely to save much more energy than carpooling and deserves serious consideration. The average van will accommodate as many passengers as at least two automobiles, and the energy costs of operating a van are often less than those of one large automobile. For example, both Chevrolet and Ford made 1976 vans with EPA combined fuel economy ratings of nineteen miles per gallon, while standard size automobiles for these two companies were rated at only thirteen and fifteen miles per gallon [35]. Several industries initiated vanpooling programs among their employees by providing them with vans [36]. Where the distribution of employee residences lends itself to vanpools, governments should consider establishing such a program. Reimbursement of government costs could come through fares paid by riders. While use of an employee as driver is a possibility, some problems have arisen in this connection [37].

A CASE STUDY IN SAVING: THE OHIO STATE UNIVERSITY ENERGY CONSERVATION PROGRAM

Ohio State University (see page 141) has developed a program aimed at improving the energy efficiency of campus buildings. The program is highly successful, having achieved energy and cost savings well beyond its initial goals.

Although a university certainly is not a government in the sense used in this book, the Ohio State program is relevant to the present inquiry for two major reasons. First, much of higher education lies within the ambit of government-provided services; and governments having a significant control over the operations of colleges and universities may make direct use of information referred to in this chapter. Second, a cluster of government buildings is physically analogous to a college campus and presents many of the same problems and opportunities for energy conservation. In addition, the program at Ohio State is a recent one, begun in early 1973, and is thoroughly documented from then up to the present.

In the late sixties and early seventies, Ohio State's utility bill was rising at an annual rate of 15 percent, most of the increase due to expansion of facilities. By early 1973, the bill stood at $5 million a year [38]. Even before the energy embargo of 1973, the university anticipated increased utility rates and established an Energy Conservation Coordinating Committee to develop and coordinate energy conservation on campus. Many of the committee's conservation goals have been met without significant curtailment of operations, an achievement most impressive in view of the special attention given to environmental concerns: "The environmental needs of the operations performed in the building are evaluated, and system modifications are proposed within the limits set by those requirements" [39]. Interestingly, the committee consists entirely of faculty members and physical plant facility managers. This approach is in accord with recommendations that employee expertise and ideas be used to stimulate employee cooperation (see page 148).

In the area of management, strategies used by the committee have included the following:

- Placing advertisements in the campus newspaper to solicit the help of faculty, staff, and students in conserving energy and to inform them of progress on the campus (a strategy analogous to the use of memoranda and posters in offices to enhance employee awareness of conservation programs).

- Scheduling air conditioner spring startup and fall shutdown according to occupancy and other characteristics of individual buildings.
- Establishing standard room thermostat settings for summer and winter operation.
- Developing building lighting level guidelines designed to reduce electricity consumption.
- Starting summer office hours a half hour early so that cooling systems can be shut down earlier in the day.
- Establishing a campus control center to monitor and regulate building systems.
- Conducting classes for maintenance personnel to help them operate building systems correctly and economically.
- Controlling heating and ventilating systems and exhaust fans both by hand and with time clocks (this method, partly a retrofitting strategy, has saved up to 30 percent of original utility costs).
- Reducing exhaust and ventilation air requirements.
- Scheduling building hours to minimize the number of buildings operated during night, weekend, and holiday periods.

The Ohio State program has led to reductions in consumption of 16 percent for natural gas and fuel oil, and 14 percent for electricity. Dollar savings in fiscal year 1974—1975 totaled over $1.2 million, with $2.4 million the projected savings for 1975—1976.

Building modification, or retrofitting, is a major part of the Ohio State program. By the summer of 1976, six buildings had been modified, with projected energy savings of 25 percent to 50 percent for each building. University officials emphasize the need for professional evaluation on a building-by-building basis: "There is no simple list of rules that can be universally applied to saving energy in buildings. Each system must be evaluated on an individual basis in order to achieve maximum operating efficiencies" [40]. Table 7—2 lists, for each of the six buildings, the costs of retrofitting and the resulting savings through reduced energy usage.

The first building modified, Allied Medical, is a five story building of offices, laboratories, and classrooms with a two story wing used for audio and television studios [41]. Floor space totals 93,000 square feet.

The premodification analysis was done in three stages:

1. Building construction specifications were examined to determine energy usage levels of heating, cooling, lighting, and ventilating systems, and to check for any other energy uses anticipated at the

Table 7-2. Six Retrofitted Buildings—Costs and Savings

Ohio State University Conservation Program

Building	Cost of One Time Modification	Annual Savings *	Number of Months Monitored
Allied Medical	$ 30,000	$ 61,756	15
Health Science Library**	27,500	69,830	9
Electronics Laboratory**	25,750	41,935	6
Biological Sciences	37,156	81,765	9
McCampbell Hall	54,200	60,820	12
Howlett Hall	31,738	52,230	9
	$206,344	$368,336	

*Based on monitored energy use over the number of months shown and projected to total yearly savings at current utility rates.

**Savings based on estimates of steam costs during a period just prior to modifications, when no steam-metered records were available. Steam meters are now recording building consumption. Electricity was metered both before and after modification.

Source: Robert H. Fuller, Dallas Sullivan, and Charles F. Sepsy, "The Ohio State University Energy Conservation Program: Methods and Results" (Presented at the Annual Meeting of the American Society of Heating, Refrigerating and Air-Conditioning Engineers, Seattle, Washington, June 27, 1976).

time the building was constructed. The existing systems were then observed and tested to determine actual performance and energy consumption. Lighting and other energy usage patterns were noted on a room-by-room basis.

2. Energy usage was categorized according to function (heating, lighting, elevators, and so on).

3. Using the information gathered, modifications were proposed for each building and its systems.

Consumption for the fiscal year 1972–1973 was determined to be 2,742,683 kwh (electricity) and 34,974 mcf (natural gas), which works out to 478,000 BTU a year for each square foot of floor space. Costs for this energy totaled $102,400.

Modifications made to the Allied Medical building were as follows:

1. Ventilating systems were shut down when the building was not occupied.

2. Ventilation was reduced in the audio/television wing through the installation and use of new sheaves and two speed motors with maximum capacities lower than for equipment used previously.

3. Twenty-two percent of fluorescent tubes were removed, resulting in a reduced cooling and fan system load.
4. The hot water supply temperature for winter was reduced from 240° F to 190° F.
5. Air handling unit mixed air setting was increased from 55° F to 63° F during the winter months.
6. Thermostats were set at 68° F for winter and 76° F for summer (with the exception that variable volume units, which are not capable of heating the space, remained constant at 76° F).
7. The standby boiler was shut down.
8. The hot deck heating coil control was improved by installation of a three way valve.

These modifications cost $30,000, and the payoff was dramatic in terms of both energy and money. For the 1975 calendar year, electricity consumption was 1,554,406 kwh and gas consumption was 7566 mcf—lower by 43.3 percent and 78.4 percent, respectively, than the figures for fiscal year 1972–1973. The cost savings attributable to these reductions in consumption was $61,756—more than double the cost of the retrofitting within a single year! How else is it possible to earn an annual return of over 108 percent, with the entire initial investment returned within about six months, at virtually no risk?

How efficient was the Allied Medical building after it was modified? The answer is, of course, relative. A good measure of efficiency is the energy performance index (EPI), defined as the energy usage in BTUs per square foot of floor space per year. The EPI for Allied Medical building fell from 478,000 to 138,750. Although this reduction is certainly dramatic, the Allied Medical building as modified remains relatively inefficient. The U.S. General Services Administration has set an EPI target of 75,000 for its retrofitted office buildings [42]. The discrepancy is explained in Senate testimony by members of the Ohio State engineering faculty:

> Other more costly modifications such as heat recovery units for buildings having high exhaust requirements, building insulation, and replacement of heat-driven absorption chilling units with electrically-driven units have been evaluated and, at current utility rates, show a four to eight year payback. *These are being deferred to fund the more productive modifications* [43].

With 360 buildings and 16,000 employees, Ohio State does not have the capital to immediately carry out every modification justifiable over the long term. Accordingly, they are applying available capital where it will yield the highest returns. The Ohio State program clearly demonstrates that retrofitting is no temporary strategy. The Energy Conservation Coordinating Committee is a permanent institution, continually evaluating and balancing factors such as energy costs, available technology, and building utilization in order to schedule further modifications.

This example of the dramatic savings to be achieved through changes in routine and structure ends Chapter 7 on an upbeat. Readers interested in the Ohio State program should contact Robert H. Fuller, P.E., Department of Mechanical Engineering, 206 West 18th Avenue, Columbus, Ohio 43210.

NOTES TO CHAPTER 7

1. Categorization should be on the basis of (1) duration of item's usefulness, (2) efficiency of the item relative to that of possible replacements, (3) efficiency of the item relative to what could be achieved by modification, (4) cost of replacement, and (5) cost of modification.

2. This is essentially the approach to "energy audits" taken in the "supplemental state energy conservation plan" provisions of the recently enacted Energy Conservation and Production Act, §§431–32, Pub. L. No. 94–385, 90 Stat. 1125 et seq. (Aug. 14, 1976), *amending* Part C of Title 3 of the Energy Policy and Conservation Act, §§361–66, Pub. L. No. 94–163, 42 U.S.C. §§6321–26 (Dec. 22, 1976). See Chapter 6 for a brief discussion of these plans.

3. *See generally* Denis Hayes, *Energy: The Case for Conservation* (Washington, D.C.: Worldwatch Institute, 1976).

4. Center for Science in the Public Interest (CSPI), *People and Energy* vol. II, no. 5 (Washington, D.C.: CSPI, May 1976), p. 4.

5. This discussion is substantially derived from Federal Energy Administration (FEA), Office of Energy Conservation and Environment, *Guidelines for Saving Energy in Existing Buildings: Engineers, Architects, and Operators Manual*, ECM 2, Conservation Paper No. 21 (Washington, D.C.: FEA, June 1975), pp. 38–150.

6. *Ibid.*, pp. 38–40.

7. *Ibid.*, p. 40. Costs in individual circumstances may vary from this generalized comparison.

8. *Ibid.*, pp. 49–57.

9. *Ibid.*, p. 58.

10. *Ibid.*, pp. 58–65.

11. *Ibid.*, p. 58.

12. *Ibid.*, p. 38; and U.S. General Services Administration (GSA), *Energy Conservation Guidelines for Existing Office Buildings* (Washington, D.C.: GSA, 1975), pp. 3–6.

13. FEA, *supra* note 5 at 38.

14. GSA, *supra* note 12 at 3–6.

15. FEA, *supra* note 5 at 38.

16. GSA, *supra* note 12 at 3–11.

17. *See generally*, Belinda L. Collins, *Windows and People: A Literature Survey: Psychological Reaction to Environments With and Without Windows* (Washington, D.C.: National Bureau of Standards, Building Science Series 70, 1975).

18. FEA, *supra* note 5 at 34.

19. GSA, *supra* note 12 at 3–11.

20. FEA, *supra* note 5 at 28.

21. Kneeland A. Godfrey, Jr., "Energy Conservation in Existing Buildings," *Civil Engineering—ASCE*, September 1975, pp. 81–82.

22. This difficult concept is discussed further in Frederick J. Wells' book of this series, *Utility Pricing and Planning: An Economic Analysis* (Cambridge, Massachusetts: Ballinger, 1977).

23. *See* Bureau of National Affairs, *Energy Users Report*, January 8, 1976, pp. D–7—D–10.

Note also that the District of Columbia has installed load management systems in several district facilities, and has realized savings of up to 30.5 percent in energy usage. Information courtesy of Mr. Malcolm Clark, chief, Improvement, Maintenance and Planning Division, Bureau of Repairs and Improvements, District of Columbia Government.

24. For information on IBM's System/7 Power Management System, call Jack Brock, Ron Kath, or George Magner at (404) 252–6820, or write to them at IBM General Systems Division, 5775–D Glenridge Road Northeast, Atlanta, GA. 30328.

25. The major national regulations governing the health and safety of the workplace environment are promulgated by the United States Occupational Safety and Health Administration, pursuant to the Occupational Safety and Health Act of 1970, 84 Stat. 1590, 29 U.S.C. §§651 et seq. These regulations are *not* applicable to state and local governments, because the act defines "employer" so as not to include "any State or political subdivision of a State." 29 U.S.C. §652(5). When individual states take over enforcement of the act, however, as twenty-three have done so far, then their regulatory schemes must apply to the state's own agencies. 29 U.S.C. §667(c) (6).

26. Federal Energy Administration (FEA), *Energy Conservation Site Visit Report: Toward More Effective Energy Management* (Washington, D.C.: GPO, April 1976) (available from the Superintendent of Documents, U.S. Government Printing Office, Washington, D.C. 20402, Stock No. 041–018–00100–0, for $1.70).

27. *See*, United States General Accounting Offices (GAO), *Report to the Congress: Energy Conservation at Government Field Installations—Progress and Problems* (Washington, D.C.: GAO, 1976).

28. *Ibid.*, pp. 3–4.

29. Note that this subject is treated comprehensively in Durwood J. Zaelke, *Urban Transportation and Energy Conservation* (Cambridge, Massachusetts: Ballinger, 1977).

30. *See, e.g.*, Cal. Gov't Code §14622 (West Supp. 1976), allowing the director of the department of general services to provide parking for elected officials without charge.

31. For an excellent discussion of the potential impact of various parking management programs (e.g., parking taxes, rate regulation, parking supply restrictions, etc.) on the number of auto-driving commuters and vehicle miles traveled, *see, Parking Management Policies and Auto Control Zones*, Metropolitan Washington Council of Governments, Summary Report prepared for U.S. Department of Transportation (February 1976).

32. FEA, *supra* note 2 at 30.

33. *Ibid.*

34. 41 Fed. Reg. 7944 (February 23, 1976).

35. U.S. Environmental Protection Agency (EPA), *1976 Gas Mileage Guide for New Car Buyers* (Washington, D.C., 1975).

36. *See e.g.*, Robert D. Owens and Helen L. Sever, *The 3M Commute-A-Van Program: Status Report* (St. Paul, Minnesota: 3M Co., 1974).

37. *See* Zaelke, *supra* note 5.

38. This description is substantially derived from the statement of Thomas B. Smith, P.E., and Robert H. Fuller, P.E., before the Senate Interior Committee on April 26, 1976.

39. *Ibid.*, p. 4.

40. Robert H. Fuller, P.E.; Dallas Sullivan, P.E.; and Charles F. Sepsy, P.E., "The Ohio State University Energy Conservation Program: Methods and Results" (Presented at the Annual Meeting of the American Society of Heating, Refrigerating, and Air-Conditioning Engineers, Inc., in Seattle, Washington, June 27, 1976), p. 14.

41. This description of retrofitting the Allied Medical Building is substantially derived from the presentation of Fuller, Sullivan, and Sepsy, *Ibid.*, pp. 15–23.

42. United States General Services Administration (GSA), *Energy Conservation Guidelines for Existing Office Buildings* (Washington, D.C., 1975), pp. 1–5.

43. Statement of Smith and Fuller, *supra* note 38, p. 6. Emphasis added.

Appendices to Chapter 7

The bibliography below contains sources of methods which can be applied by governments and their agencies in establishing programs to conserve energy.

Carrillo, Hernando (planning analyst, state of Florida, department of general services). *Discussion Paper on: Staggered Work Hours for the Larson Building.* Tallahassee, Florida: Department of General Services, 1973.

Federal Energy Administration. *Energy Conservation Site Visit Report: Toward More Effective Energy Management.* FEA Conservation Paper Number 38. Washington, D.C.: GPO, 1976. (For sale by the superintendent of documents, U.S. Government Printing Office, Washington, D.C. 20402—Price $1.70. Stock Number 041—018—00100—0.)

Federal Energy Administration. *Guidelines for Saving Energy in Existing Buildings: Building Owners and Operators Manual.* FEA Conservation Paper Number 20. Washington, D.C., 1975.

Public Technology, Inc. *Energy Conservation: A Technical Guide for State and Local Governments.* Washington, D.C., 1975. (Contact: Technology Exchange Program, Public Technology, Inc., 1140 Connecticut Avenue, N.W., Washington, D.C. 20036. Telephone: (202) 223—8240.)

State of Oregon, Department of Energy. *Energy Todate*, July 1976 and August 1976. Salem, Oregon, 1976. (See the articles on the SHUTTLEBUG van service. Contact: Ms. Zoe A. Wilson, Editor, Department of Energy, 528 Cottage Street N.E., Salem, Oregon 97310. Telephone: (503) 378—4129.)

Tennessee Energy Office. *Recommendations for Greater Energy Efficiency in Large Buildings.* Nashville, Tennessee. (Contact: Tennessee Energy Office, 250 Capitol Hill Building, Nashville, Tenn. 37219. Telephone: (615) 741—1772.)

United States General Accounting Office. *Report to the Congress: Energy Conservation at Government Field Installations—Progress and Problems.* Washington, D.C., 1976. (The cost of this report is $1.00 per copy. Officials of federal, state, and local governments may receive up to ten copies free of charge. Members of the press, college libraries, faculty members, students, and nonprofit organizations may receive up to two copies free of charge. Order Number and Date: LCD−76−229; Aug. 19, 1976. Those entitled to free copies, contact: U.S. General Accounting Office, Distribution Section, Room 4522, 441 G Street, N.W., Washington, D.C. 20548. Those who are required to pay, contact: U.S. General Accounting Office, Distribution Section, P.O. Box 1020, Washington, D.C. 20013. Please do not send cash.)

United States General Services Administration. *Energy Conservation Guidelines for Existing Office Buildings.* Washington, D.C., 1975.

United States General Services Administration. "Management of Buildings and Grounds: Conservation of Energy in Government-Owned and Government-Leased Space. 41 C.F.R. 101−20.116, 39 Fed. Reg. 39266 (Nov. 6, 1974), ELR 46801.

United States National Bureau of Standards. *Government Activities and Regulations for Buildings on Energy Saving Standards: A Look at What Various Agencies Have Done and Intend to Do for Energy Conservation.* Washington, D.C., 1975. (Available for $3.50 from the National Technical Information Service, U.S. Department of Commerce, Springfield, Virginia 22161. Also published in *Heating/Piping/Air Conditioning* 47, no. 13 (December 1975): 41−46.)

Ventura County Superintendent of Schools. *The Energy Crisis in the Public Schools: Alternative Solutions.* Ventura, California, 1974. (Contact: Dr. James F. Cowan, Superintendent, Ventura County Superintendent of Schools, County Office Building, Ventura, California 93001.)

Young, Dr. John C., in Collaboration with Bruce C. Wyman and Cherry K. Owen, Lake Charles−McNeese Urban Observatory, Inc. *Energy Use, Energy Savings and Cost Reductions in All Operating Departments City of Lake Charles.* Lake Charles, Louisiana, 1976.

GENERAL SERVICES ADMINISTRATION REGULATIONS FOR CARPOOL PARKING, 41 FED. REG. 7944−7945, 41 C.F.R. PART 101−20 (FEBRUARY 10, 1976) [FPMR Amendment D−52]

Part 101−20 —Management of Buildings and Grounds

Carpool Parking

This regulation provides policies and procedures related to employee carpool parking.

1. The table of contents for Part 101−20 is amended to provide new entries as follows:

101−20.117   Carpool parking.
101−20.117−1   Definitions.
101−20.117−2   Policies.
101−20.117−3   Leased or contractor-operated parking spaces.
101−20.117−4   Guidelines for implementation.

Subpart 101–20.1–Building Operations, Maintenance, Protection, and Alterations

2. Section 101–20.117 is added as follows:

§101–20.117 Carpool Parking

§101–20.117–1 Definitions

The following definitions shall apply to this section:

(a) "Agency parking" means vehicle parking spaces under the jurisdiction and/or control of a Federal agency which are used for parking Government vehicles, other official vehicles, visitor vehicles, and employee vehicles.

(b) "Carpool" means a group of two or more people using a motor vehicle for transportation to and from work.

(c) "Employee parking" means the parking space assigned for the use of employee-owned vehicles other than those classified as "official parking."

(d) "Federal agency" means any executive department or independent establishment in the executive branch of Government, including any wholly owned Government corporation.

(e) "Handicapped employees" means Government employees so severely physically handicapped as to prohibit or make unreasonably difficult the use of public transportation. Justification for this priority may require certification by an agency medical unit or the Public Health Service.

(f) "Official parking" means parking spaces reserved for Government-owned, Government-leased, or privately owned vehicles regularly used for Government business. The phrase "privately owned vehicles regularly used for Government business" means vehicles used 12 or more workdays per month for Government business for which the employee receives reimbursement for mileage and parking fees under Government travel regulations. Monthly certification by agency heads may be required to establish this priority.

(g) "Parking space" means the area allocated in a parking facility for the temporary storage of one passenger-carrying motor vehicle.

(h) "Regular member" means a person who travels daily (leave excepted) in a carpool for a minimum distance of 1 mile each way. In addition, an agency may define a regular member as one whose worksite is located within a specific but reasonable distance from the parking facility.

(i) "Visitor parking" means parking spaces reserved for the exclusive use of visitors to Federal facilities.

§101–20.117–2 Policies

Agencies shall encourage the conservation of energy by taking positive action to increase carpooling. The following policies shall be reflected in agency plans:

(a) *Parking.* In assigning all parking spaces assigned to or controlled by each agency, the following policies shall be observed:

(1) Agencies shall give first priority to official and visitor parking requirements.

(2) Severely handicapped Government employees for whom assigned parking spaces are necessary shall be accommodated.

(3) A goal of not more than 10 percent of the total spaces available for employee parking on an agency-wide basis (excluding spaces assigned to severely handicapped) shall be assigned to executive personnel and persons who are assigned unusual hours.

(4) All other spaces available for employee parking shall be made available to carpools to the extent practical.

(5) Those parking spaces reserved for carpools shall be assigned primarily on the basis of the number of members in a carpool.

(6) For the purpose of allocation of parking spaces for carpools, full credit shall be given to any regular member regardless of where he is employed except that at least one member of the carpool must be a full-time employee of the agency.

(b) *Two-wheeled vehicles.* Subject to the availability of satisfactory and secure space and facilities, agencies shall reserve areas for the parking of bicycle and other two-wheeled vehicles. Bicycles shall be given special consideration including storage type space in buildings and improved bicycle locking devices where practical and appropriate funds are available. Bicyles shall not be transported on elevators or via stairways, or parked in offices.

(c) *Regular hours.* Agency managers and supervisors shall make every effort to maintain regular arrival and departure times for all employees. Supervisors are reminded of their prerogative, within overall agency policy, to adjust the scheduled duty hours of individual employees to facilitate carpooling and the use of mass transit.

§101–20.117–3 Leased or Contractor-Operated Parking Spaces

When parking spaces are controlled by specific lease or other contractual agreements, appropriate agency contracting officers shall endeavor to amend the contracts to the extent necessary to accomplish the policies prescribed by this regulation, provided the amendments are not otherwise adverse to the best interests of the Government. Where it is not economically prudent to amend existing contracts, the contracts shall be modified before renewal to comply with the agencies prescribed parking procedures.

§101–20.117–4 Guidelines for Implementation

Agencies shall develop and implement employee carpooling programs through extensive promotional campaigns using available internal communications. Agencies shall be responsible for assigning employee parking spaces assigned to or under the control of that agency. Implementation of the provisions of this regulation may require consultation, as appropriate, with recognized labor organizations. Each agency shall maintain written plans and procedures for the assignment of parking spaces including as a minimum the following items:

(a) Specific methods and procedures to be followed by the agency in the assignment of employee parking spaces;

(b) Assistance available to employees in establishing or joining carpools and the procedures to be followed in filing applications for parking spaces;

(c) Provision for at least an annual review and reassignment of all parking spaces;

(d) Procedures for interim reassignment and replacement caused by membership turnover;

(e) A definition of employee responsibility in the use of the parking spaces and in promptly reporting any changes in the number in or membership of carpools;

(f) A statement of penalties for misrepresentation of carpool applications (A mandatory penalty of at least 6 months' suspension of the privilege of parking on a Federal facility shall be imposed for misrepresentation of carpooling membership, application qualifications, or for violation of other agency carpooling practices and requirements. The agency may also impose other penalties where appropriate.);

(g) Provision for enforcing the parking rules and regulations;

(h) A system for maintaining carpool records and files; and

(i) Provision for internal monitoring and auditing to determine compliance with these regulations.

(Sec. 205(c), 63 Stat. 390; 40 U.S.C. 486(c) and Federal Management Circular 74—1).

Effective date. This regulation is effective February 23, 1976.

Dated: February 10, 1976.

Jack Eckerd,
Administrator of General Services.

[FR Doc 76—4973 Filed 2—20—76; 8:45 am]

✳ *Chapter 8*

Disposition of Government Property

INTRODUCTION

Having procured an energy-efficient product and shep-
herded it through all stages of government ownership to
minimize its energy consumption, the energy-conscious
official is faced with a third and final problem: What is to be done
with the product, and with other products that do not consume energy, when they are no longer needed or are beyond repair? This
final question bears directly on efforts to conserve energy and other
resources. As discussed in Chapter 2, the U.S. economy has become
highly energy-intensive. Labor has been replaced by machines, and
certain primary raw materials have been replaced by synthetics. Both
shifts have frequently engendered massive increases in the energy
required for production.

A piece of paper, a metal desk, or an automobile is not often
thought of as "energy;" yet each item nonetheless *embodies* energy.
Conservation of this embodied energy is first of all dependent upon
proper treatment of the product [1]. Second, conservation goals
are served when functional items are circulated rather than "moth-
balled" when they are no longer needed. If a usable product sits in
one agency's storeroom while it is needed by another agency, the
second agency may needlessly procure an additional product. Third,
products that are no longer functional should have their usable components and materials recycled. The use of secondary, rather than primary, materials in production can save enormous amounts of energy
[2]. In addition, where component materials are combustible and

not easily recycled, they may be burned to produce energy. Governments (as well as private entities) are entrusted with the energy embodied in their material possessions; government officials have a responsibility to preserve as much of this energy as possible and to pass it along. This section suggests various means of meeting this responsibility.

It is noted with enthusiasm that strategies presented in this section would fit equally well under the perhaps more familiar rubric of "solid waste reduction" [3]. Solid waste disposal itself, not to mention the energy and natural resource waste involved, presents a substantial national problem—a problem that's increasing since waste heaps grow by 8 percent annually at the present rate [4]. In the United States, where energy appetites are enormous and where 7 percent of the world's population manages to consume one-half of the world's industrial raw materials, the "gross national product" stands at about 360 million tons of solid waste annually. This solid waste includes seventy-one billion cans, thirty-eight billion bottles and jars, four million tons of plastic, seven and six-tenths million television sets, seven million cars and trucks, and thirty-five million tons of paper [5].

How is this problem to be tackled? The U.S. Environmental Protection Agency's *Third Report to Congress: Resource Recovery and Waste Reduction* says that source separation, the principal method of resource recovery to date, is unlikely to reach its potential by 1985 and that the technology for recovery from mixed wastes is still in the development stage [6]. These twin shortcomings highlight the importance of a third approach: waste reduction. Resource *consumption* is the origin of the solid waste problem. By reducing consumption governments can drastically reduce the amount of recoverable resources that must be processed.

Many energy conservation strategies discussed in both this section and other sections of this book amount to waste reduction strategies. As discussed in Chapter 2, governments can save energy and resources, and also reduce solid waste, by purchasing durable products that use minimal quantities of resources and that come packaged in reusable containers. By promptly transferring items from agencies that no longer need them to agencies that do, or by selling such items to the public, governments can, this chapter contends, prolong the useful life of their property and thus save it from the flood tide of the solid waste stream for as long as possible.

Even a government that has exhausted its waste reduction strategies still has many options. As centralized sources of solid waste, governments are in a good position to undertake source separation programs. Unfortunately, the traditional approach to waste has been to throw it all in the same pile. But by separating waste into categories, governments can greatly facilitate the resource recovery process. These separated, homogeneous wastes can be recycled at state facilities or sold to private firms, since separation enhances salability.

Again, the strategies discussed in this chapter merit special attention since they confront simultaneously three major social challenges: (1) energy conservation; (2) natural resource conservation; and (3) solid waste disposal (which often produces air and water pollution).

TRANSFER AND SALE OF USEABLE ITEMS TO SECONDARY USERS

This part concerns government property that is no longer needed by the agency that possesses it, but which remains in useable condition. As discussed in the introductory chapter of this section, energy conservation requires that such property be transferred to another agency or to a private entity that can use it. It may be transferred either within the government, to other governments, or to the general public. The goals here are quite simple—speedy transfer of government property to entities that need it, with an intent to minimize inventory coupled with an effort to stretch each item's useful life.

The redistribution of functional items can be made part of a state or locality's inventory control procedure. This strategy, which will result in substantial dollar and energy savings, is simple and inexpensive to administer.

The first step in implementing this strategy is to get agencies to declare as *surplus* that property which is in excess of their present requirements. No matter how elaborate a redistribution procedure is devised, nothing can be done until the surplus property is identified. Failure to identify surplus promptly may result from inefficient property management at the agency level. It may also result from a deliberate effort on the part of the holding agency to maintain a stock of surplus for fear that the purchasing agency will not provide replacements promptly when they are needed. ("Better hang on to that. It took us six months to get one last time!") This hoarding, which in some instances may be essential to the smooth operation of an individual agency, is highly detrimental to government economy and energy conservation. If redistribution is to be an effective strategy, purchasing agencies and others responsible for property management must endeavor to supply agency needs with reasonable promptness. Agencies that have confidence that their future needs will be quickly met are much more likely to cooperate in relinquishing surplus.

Once surplus is identified, the purchasing department (or surplus property division—terms will vary by state and by municipality) must match it up with the parties that need it. Priority for assignment of surplus goods should be granted to the entities closest to the holding agency. For example, if a state agency made surplus available, the property would first be offered to other departments within that agency and then to other state agencies, after which it would be offered for sale to other political bodies or to the general public.

Intra-agency redistribution is the simplest form, requiring only that offices be "surplus conscious." The agency inventory control officer should keep track of property in excess of requirements. A phone call or memo may be all that is required for redistribution. Action at this level is likely to be most prevalent where storage space is in short supply [7]. No action by the purchasing department is necessary [8]. The department would simply be notified of the new location of the item after the transfer [9].

Where the transfer is between two agencies, matching up supply and demand is best accomplished by the purchasing department. Purchasing department involvement may be minimal or quite extensive [10]. Purchasing personnel can act merely as liaison between offering and requesting agencies, with little or no expansion of existing functions [11]. In this case, agencies themselves would carry out most of the details of transfer, subject to purchasing department approval.

Factors to consider in deciding whether purchasing department involvement should be at a low or high level include: (1) size and availability of work force of either the agencies or the purchasing department; (2) funding; and (3) desired scope of the redistribution program.

The extent of purchasing department involvement in the transfer process will in part determine whether any service fee is levied. For instance, the Florida Bureau of State Surplus Property, which has a relatively detailed procedure for surplus property transfers, imposes a 5 percent service fee (minimum $10.00) of the original purchase price on interagency transfers [12]. The nature of the transfer, i.e., "paper" or "cash," may also affect the question of a service fee. Ideally, in cases where there is normally a problem in persuading agencies to declare their surpluses, no service fee should be charged. Further, transfers should be made on a cash basis wherever possible, so that agencies would be encouraged to "sell" surplus items in order to obtain money to meet other needs [13].

The purchasing department should maintain a list of items requested by state agencies [14]. As requested items are declared surplus, the requesting agency would be notified immediately, and either the holding agency or the purchasing department could then proceed with the transfer.

If an item declared surplus is not intercepted by a prior request, it must be stored. Circulation of a bulletin listing available surplus property should be an effective means of disposing of warehoused property in most situations, although some states may find that turnover is too rapid to keep a bulletin current [15].

Storage may be at the holding agency or in a central location. Factors to be considered in this decision are: (1) availability of storage facilities, (2) staff size, (3) geographic size of the state, and (4) desired scope of the redistribution program. If, for example, the purchasing department's role is basically supervisory, with most of the actual transfer activity being carried out by the offering and receiving agencies, on-site storage may be preferable to central storage. On the other hand, central storage may facilitate a more active role by the purchasing department, where such is deemed desirable [16]. In some cases, expense of storage may dictate sale.

Agencies within the same jurisdiction should receive preference in the redistribution process. However, if no interagency transfer occurs within a reasonable time, then the property should be offered for sale to other political entities, nonprofit organizations, and the general public. Property can be sold by auction, sealed bid, or set price sale, depending on the nature of the item. As in the case of interagency transfers, requests may be taken; bulletins circulated; or the property may be stored in warehouses, open for inspection and sale. Generally, there must be compliance with competitive bidding laws [17].

In sum, redistribution of surplus property is a promising energy- and money-saving technique. Large-scale redistribution programs and commensurate savings are possible, but the heart of this strategy lies in overcoming hurdles of hoarding or inertia that prevent agencies from identifying their surplus and making it available to others. If a purchasing department does nothing more than prod government agencies into actually declaring unneeded property to be surplus, it will have made a significant contribution.

RESOURCE RECOVERY FROM GOVERNMENT PROPERTY THAT IS NOT IN USEABLE CONDITION

When an item of government property is either so obsolete or so badly worn as to be useless, or is simply beyond repair, it is at the junction of two divergent paths. One leads to the scrap heap; the other—too often the road not taken—leads to a recycling center. Propelling property down the recycling path guarantees substantial energy savings. A study cited in Chapter 2, for example, shows that recycling aluminum produces energy savings of up to 96 percent [18]. Numerous other such opportunities and techniques for cutting energy losses also merit governmental investigation.

Strategy: Cannibalizing

Cannibalizing, or salvaging functioning parts from irreparable items, is one way of saving objects that are in a transitional state between usable and nonusable condition from the trash heap. Since recycling processes tend to use significant energy, it is better from an energy viewpoint to reuse parts than to recycle them.

Surplus property can be cannibalized at the agency level or centrally. Sequentially, cannibalizing represents a detour in the route between the place where a machine's use is terminated and the place where the machine's component materials are recycled for secondary use. Assuming that the cannibalized parts represent a small portion of the weight of the machine, transportation costs can be saved by cannibalizing at one of these places and then shipping only the parts to central storage points. In general, the optimal location for cannibalizing may vary according to the relative locations of facilities involved, the weight and bulk of the machine, the weight and bulk of the parts to be removed, and the difficulty of removing the parts. Where possible, cannibalizing should be carried out in state-owned repair shops.

Cannibalizing cannot, of course, be economically viable unless markets for the retrieved parts exist or can be created. A single agency could conceivably provide an adequate market for its own parts, but such a marketing system is unlikely to prove feasible, and, even if it did, it might lead to improper disposal of parts that are not immediately needed. Consequently, a system for redistribution (and perhaps sale) of cannibalized parts must be set up. Ideally, this marketing network could be part of a larger system set up to redistribute agency surplus (such as the one discussed in the preceding chapter).

Strategy: Source Separation

In the words of a spokesman for the U.S. Environmental Protection Agency, source separation involves the "setting aside of recyclable materials at their point of generation by the generator" [19]. Obviously, this process is not economically feasible for all products— no one yet advocates complete categorization of all garbage! However, source separation of certain recyclable materials, notably paper, newsprint, bottles and cans, is demonstrably workable.

Removing items from the disposal process before they are mixed in with other solid waste reduces the cost of recycling [20]. State agencies can provide leadership in source separation by using the EPA guidelines described below to make it a bureaucratic procedure [21]. State and local governments may even be able to go beyond the limited source separation mandated for federal agencies under those guidelines.

The EPA regulations require source separation for recycling purposes of high grade paper—"letter head, dry copy papers, miscellaneous business forms, stationery, typing paper, tablet sheets, and computer print out paper and cards" [22] —in offices comprised of one hundred or more workers [23]. EPA recommends using the "desk-top system" of separating high grade paper. This method involves placing an extra container for recyclable paper on all desks. EPA considers the desk-top system superior to both the two basket method, which may result in too much mixing of nonrecyclables, and the centralized container system, which might inconvenience individual employees. These considerations are not so trivial as they may sound. If a system disrupts existing office routine, agency personnel at all levels will resist it.

EPA regulations also require (1) the separation of newspapers at residential facilities of more than 500 families, and (2) the separation of corrugated containers at commercial establishments generating ten or more tons of such waste per month [24]. In addition, the regulations *recommend* the separation of glass, cans, and mixed paper [25].

The EPA's limited program is probably based upon the assumption that recyclable material collected via source separation will be sold to private recycling facilities. Since the program cannot be implemented without a secure market, the government may hang back on recycling efforts. But there are two possible solutions: (1) the department of general administration (or other agency in charge of government property and operations) can construct and operate a facility to recycle government waste; or (2) the department of general administration together with the local waste disposal agency can construct and operate a facility to recycle both government and municipal (or regional) waste. These and other options are discussed below.

Strategy: Resource Recovery

Resource recovery systems process solid waste to recover natural resources or energy, or preferably *both*. As suggested above, government agencies with "custody" of government waste can (1) sell it to private facilities; (2) process it in their own facilities; or (3) coordinate with municipal or regional facilities, where they exist, to recycle it.

Selling It. While a number of governments may be able to join forces with municipal or regional recycling facilities (the third alternative), most will have to sell their recyclable waste to private recyclers. However, because governments are both major purchasers and centralized disposers, they are in a favorable position to influence the market. This stabilizing market influence can encourage recycling and can overcome one of the major impediments to recycling efforts—an unsteady market. Such uncertainty, a good example of which was the sharp fluctuation of both the supply of and the demand for waste paper during 1973–1974, thwarts investment in resource recovery facilities [26]. By supplying recyclable waste and by buying products containing recycled material (see Chapter 2), governments can play a key role in reducing market uncertainty.

Special Recovery Facilities for Government Waste. Obviously, only governments of substantial size can support resource recovery facilities on the strength of government waste alone. Furthermore, governments considering such a step should evaluate its likely effect upon the development of private facilities and upon the general market aspects of recycling. Despite these limitations, however, some large government complexes in areas with neither municipal nor private recovery facilities may find this course desirable. For one thing, it may be much more efficient to process the waste and then to transport the product to a market than to transport the waste itself to a distant processing facility.

The EPA recently issued guidelines to "provide requirements and recommended procedures for the establishment and utilization by Federal agencies of facilities to recover resources from residential, commercial, and institutional solid waste . . ." [27]. Guidelines are also intended to "recommend the establishment and utilization of such facilities to State, interstate, regional, and local governments" [28]. The guidelines do not recommend specific technologies or systems, since the EPA believes that "no one system is best under all circumstances" [29], but they do require that a federal facility that "generates, collects, or disposes of 100 tons or more per day of solid waste" must "establish and/or utilize" recovery facilities [30]. There is also a provision setting the same requirement where the aggregate waste of several federal facilities in a single standard metropolitan statistical area (SMSA) equals one hundred tons or more [31]. Reproduced in the appendix, the EPA guidelines present a valuable model for state and local efforts.

Coordination With Municipal or Regional Facilities. A number of localities have constructed, are constructing, or plan to construct facilities for recovery of energy and other natural resources from solid waste [32]. When the economic value of recovered resources is coupled with local savings for solid waste management, these facilities can actually generate revenue for the jurisdiction [33]. Consequently, this third and final option, coordination of government waste disposal with that of the surrounding region, is the best and most efficient—when available.

For further information on municipal recycling and its integration with preexisting solid waste management, see Norman L. Dean's book of this series, *Energy Efficiency in Industry: A Guide to Legal Barriers and Opportunities* [34]. The primary point to be made here is that governments can provide solid waste to such facilities and can also stimulate the purchase of recycled material.

The Resource Conservation and Recovery Act of 1976: New Federal Incentives for State and Regional Action

In the fall of 1976, not long before this book went to press, Congress enacted legislation amending the Solid Waste Disposal Act [35]. It appears that this somewhat complex act will have a significant effect upon resource recovery at the state level. Subtitle D of the act authorizes the appropriation of $70 million to assist states or regional authorities to develop and implement solid waste management plans [36]. Authorizations are for $30 million to be appropriated for fiscal year 1978 and $40 million for fiscal year 1979 [37].

Assistance is subject to many exceptions and conditions, so that complete understanding requires a very careful reading of the act, but the basic purposes to which financial assistance can be applied are as follows:

Such assistance shall include assistance for facility planning and feasibility studies; expert consultation; surveys and analyses of market needs; marketing of recovered resources; technology assessments; legal expenses; construction feasibility studies; source separation projects; and fiscal or economic investigations or studies; but such assistance shall not include any other element of construction, or any acquisition of land or interest in land, or any subsidy for the price of recovered resources [38].

In order to be eligible for federal assistance, states must develop plans. Primary requirements for these plans include:

1. Identification of (a) the breakdown of implementation responsibilities among state, local, and regional authorities; (b) the distribution of federal funds to these authorities; and (c) the means for coordinating regional planning and implementation under the plan [39].
2. Prohibition of establishing new open dumps [40].
3. The requirement that all solid waste, except hazardous waste, be utilized for resource recovery, disposed of in sanitary landfills, or otherwise disposed of in an environmentally sound manner [41].
4. Provision for the closing or upgrading of all existing open dumps within the state [42].
5. Establishment of such state regulatory powers as may be necessary to implement the plan [43].
6. Removal of any existing prohibitions on local governments entering into long term contracts for the supply of solid waste to resource recovery facilities [44].
7. Provision that solid waste be disposed of in an environmentally sound manner [45].

These are not the sole requirements, and again the scheme is complicated, calling for close reading and interpretation of the act.

While the effect of this act upon state resource recovery may be substantial, this effect may not be felt for several years. State and local governments are accordingly urged not to retard recovery plans that are of purely local origin, but rather to go forward with such plans with an awareness of the new federal act. The act does not affect existing solid waste management or planning unless such activities "are inconsistent with a State plan approved by the Administrator ... " [46].

NOTES TO CHAPTER 8

1. *See* Chapter 4 for a discussion of one way to assess the energy embodied in various goods and services.
2. *See* Chapter 2 for examples.
3. The term solid waste refers to

> garbage, refuse, sludge, and other discarded solid materials, including solid waste materials resulting from industrial, commercial, and agricultural operations, and from community activities, but does not include solids or dissolved materials in domestic sewage or other significant pollutants in water resources, such as silt, dissolved or sus-

pended solids in industrial wastewater effluents, dissolved materials in irrigation return flows or other common water pollutants.
41 Fed. Reg. 16953, April 23, 1976.

4. H.R. Rep. No. 94–1319, 94th Cong., 2d sess., pp. 3 and 7 (1976); and U.S. Environmental Protection Agency (EPA), *Third Report to Congress: Resource Recovery and Waste Reduction* (Washington, D.C.: 1975), pp. 5 and 7.

5. H.R. Rep. No. 94–1319, 94th Cong., 2d sess., pp. 6–7.

6. EPA, *supra* note 4 at ix.

7. Perhaps this is an argument for limiting office storage space.

8. For example, Florida, which has a relatively detailed certification procedure for surplus property transfers, nonetheless permits intradepartmental transfers without involvement of the bureau of state surplus property. Florida Department of General Services, Division of Surplus Property, Bureau of State Surplus Property, Regulation No. 13F–1.10 (1973).

9. *See e.g.*, Physical Inventory Manual, State of Maryland, Department of General Services, Purchasing Bureau, July 1975, IV–1. Telephone interview with Mr. Stanley Hanna, purchasing bureau, department of general services, state of Maryland, June 14, 1976.

10. Washington, one of the first states to set up a disposition program, appears to provide for the most extensive department participation of the states surveyed (California, Florida, Maryland, New Jersey, Oregon, Pennsylvania, Washington, and Wisconsin). Telephone interview with Mr. James Hackett, state property disposal office, department of purchasing, general services department, June 16, 1976.

11. The Maryland Purchasing Bureau administers the disposition program with its regular staff. The bureau makes use of the inventory control specialist who is already required to keep an inventory summary. Agencies store their own personal property. Telephone interview with Mr. Stanley Hanna, *supra* note 3. The New Jersey program is also run by the purchasing staff. Telephone interview with Mr. Osborn, division of purchase and property, purchase bureau, state of New Jersey, June 17, 1976.

12. Florida Department of General Services, *supra* note 8.

13. In Washington, for example, vehicles are purchased through a revolving fund. Cash from sales of vehicles is returned to the holding agency to replenish its fund. Telephone interview with Mr. James Hackett, *supra* note 10.

14. The Federal Procurement Regulations take a similar stance on this problem:

> Before taking procurement action in accordance with this chapter, agencies shall have complied with applicable laws and regulations relative to obtaining supplies or services from Government sources and from contracts of other Government agencies. These include excess and surplus stocks in the hands of any Government agency, Federal Supply Schedules, General Service Administration Stores Stock, Federal Supply Sources, Consolidated Purchase Programs, Federal Prison Industries, Inc., and National Industries for the Blind.

41 C.F.R. 7–1.302–1(a).

15. Officials from Pennsylvania and Wisconsin indicated that they seldom, if ever, circulate bulletins due to the volume of turnover. Telephone interview with Mr. Leo Haley, Pennsylvania Department of General Services, Division of Surplus State Property, and Mr. Schultz, Wisconsin Department of Administration, State Purchasing Office, June 16, 1976.

16. The Pennsylvania Division of Surplus State Property operates a central warehouse in Harrisburg which is run like a retail store. Telephone interview, Mr. Leo Haley, *supra* note 15.

17. State agencies in New Jersey must get three competitive bids to dispose of their own property. Telephone interview, Mr. Osborn, *supra* note 11. When selling their own surplus property, Florida state agencies must sell to the highest bidder. A copy of the advertisement for bidding and of the bid must be sent to the Division of Surplus Property. Telephone interview with Mr. R.C. Covington, director, Florida Department of General Services, Division of Surplus Property, June 15, 1976.

18. William E. Franklin, David Bendersky, William R. Park, and Robert G. Hunt, "Potential Energy Conservation from Recycling Metals in Urban Solid Waste," in *The Energy Conservation Papers*, ed. Robert H. Williams (Cambridge, Massachusetts: Ballinger, 1975), p. 172.

19. 41 Fed. Reg. 16953 (April 23, 1976).

20. *See* Erica Dolgin and Thomas Guilbert, *Federal Environmental Law* (St. Paul, Minnesota: West, 1974), p. 1315.

21. 41 Fed. Reg. 16950 (April 23, 1976).

22. *Ibid.* at 16953.

23. *Ibid.* at 16954.

24. *Ibid.* at 16955.

25. *Ibid.* at 16954—55.

26. *See* U.S. Environmental Protection Agency (EPA), *Third Report to Congress: Resource Recovery and Waste Reduction* (Washington, D.C.: Government Printing Office, 1975), pp. 46—52.

27. 41 Fed. Reg. 41208 (Sept. 21, 1976), 40 C.F.R. Part 245.

28. 41 Fed. Reg. 41208.

29. *Ibid.* at 41209.

30. *Ibid.* at 41209—10, 40 C.F.R. §245.200—1(a).

31. 41 Fed. Reg. 41210—11, 40 C.F.R. §245.200—1(b).

32. *See e.g.*, EPA, *supra* note 9 at 87—96.

33. *See* H.R. Rep. No. 94—1319, 94th Cong., 2d sess., pp. 22—24 (1976).

34. Norman L. Dean, The *Environmental Law Institute State and Local Energy Conservation Project, Energy Efficiency in Industry: A Guide to Legal Barriers and Opportunities* (Cambridge, Massachusetts: Ballinger, 1977).

35. The Resource Conservation and Recovery Act of 1976, Pub. L. No. 94—580 (Oct. 21, 1976) *amending* the Solid Waste Disposal Act of 1965, Pub. L. No. 89—272, *as amended* 42 U.S.C. §§3251—59 (1970), ELR 41901.

36. Pub. L. No. 94—580 §§4001—09.

37. *Ibid.* at §4008(a)(1).

38. *Ibid.* at §4008(a)(2)(A).

39. *Ibid.* at §4003(1).
40. *Ibid.* at §4003(2).
41. *Ibid.* at §4003(2).
42. *Ibid.* at §4003(3).
43. *Ibid.* at §4003(4).
44. *Ibid.* at §4003(5).
45. *Ibid.* at §4003(6).
46. *Ibid.* at §4007(c).

Appendix to Chapter 8

RESOURCE RECOVERY FACILITIES GUIDELINES
DEVELOPED BY THE UNITED STATES ENVIRON-
MENTAL PROTECTION AGENCY

Title 40—Protection of Environment
Chapter 1—Environmental Protection Agency
[FRL 603-2]

Part 245—Promulgation Resource Recovery
Facilities Guidelines

Background

On January 15, 1976 notice was published in the Federal Register (41 FR 2359) proposing regulations to establish a new part 245 of Chapter I of Title 40 of the Code of Federal Regulations pursuant to the authority of Section 209(a) of the Solid Waste Disposal Act of 1965 (Pub. L. 89—272), as amended by the Resource Recovery Act of 1970 (Pub. L. 91—512). Section 209 of the amended Act requires the Administrator of the U.S. Environmental Protection Agency (EPA) to "recommend to appropriate agencies and publish in the Federal Register guidelines for solid waste recovery, collection, separation, and disposal systems (including systems for private use), * * * " In addition, section 211 mandates that Federal agencies having jurisdiction over solid waste disposal activities "shall insure compliance with the guidelines recommended under section 209 and the purpose of [the Solid Waste Disposal Act]. * * * "

In fulfillment of its responsibilities under section 209, EPA promulgated the first set of guidelines: "Guidelines for the Thermal Processing and Land Disposal of Solid Wastes," on August 14, 1974 (40 CFR 240 and 241). Since that time, guidelines have been promulgated for the Storage and Collection of Residential, Commercial, and Institutional Solid Waste on February 13, 1976 (40 CFR 243), and for Source Separation for Material Recovery on April 23, 1976 (40 CFR 246) and published in proposed form for public comment in the Federal Register on November 13, 1975 for "Beverage Containers" (40 FR 52967). In addition, non-mandatory guidelines for "Procurement of Products that Contain Recycled Material" were published in the Federal Register on January 15, 1976 (40 CFR 247).

The promulgation of these guidelines will meet the Administrator's initial obligation under Section 209 to publish guidelines in the area of solid waste recovery and separation systems. The EPA intends to revise and supplement these guidelines in the future, in recognition of the specific statutory language of section 209 that guidelines "shall be revised from time to time."

Section 211 of the Act and Executive Order 11752 make the "Requirements" section of the guidelines mandatory upon Federal agencies. The recommendatory sections of the guidelines present methods and techniques which amplify or supplement the mandatory requirements. These sections contain desirable, but not essential, procedures useful for attaining the purposes of the Act.

As provided in section 211 of the Act and section 3(a) of Executive Order 11752, heads of Federal agencies are ultimately responsible for determining which facilities under their jurisdiction shall comply with the guidelines. Pursuant to its authority in section 3(d) of EO 11752, EPA has required that each decision not to establish or utilize a resource recovery facility must be justified in a report to the Administrator and the public. The specific requirements for this report may be found in the "Scope" section of the guideline.

Since each Federal agency carries out a wide variety of activities, each with associated policy and economic considerations, EPA cannot require agencies to carry out specific actions. Rather EPA has set objectives in the guidelines, allowing each agency to determine for itself the means of accomplishing these objectives by utilizing information provided in the recommended procedures and bibliography and by utilizing technical assistance provided by the EPA.

The economic and inflationary impact of these guidelines has been carefully evaluated. It has been determined that the effects will be minor and that these guidelines are not a "major action" requiring an inflation impact statement as prescribed by Executive Order 11821 and OMB Circular A-107.

Introduction

These guidelines provide requirements and recommended procedures for the establishment and utilization by Federal agencies of facilities to recover resources from residential, commercial, or institutional solid wastes, and recommend the establishment and utilization of such facilities to State, interstate, regional, and local governments. Utilization of resource recovery facilities will result in conservation of resources and in a reduction in the amount of solid waste that requires disposal.

Public Comment

Written comments on the proposed guidelines were invited and were received from 21 sources. The commenters addressed 69 issues which fell into 21 categories. As a result of these written comments and in an effort to clarify the guidelines, certain changes were made. All of the written comments and the Agency's disposition of each comment are on file with the Agency and are available for public examination at the EPA Public Information Reference Unit (EPA Library), 401 M Street, SW., Washington, D.C. during normal business hours. The major issues raised by the commenters and the Agency's consideration of them are described below.

Compatibility, Technology and Cost. 25 comments were received concerning the compatibility of the resource recovery facilities guidelines with waste reduction and source separation techniques, and questioning the cost and availability of technology for resource recovery facilities. Technology, cost and compatibility are interdependent and must be considered as interrelated parts of a solid waste management system. They must also be examined in light of the Congressional mandate for solid waste management by Federal agencies. The Congress intended that Federal agencies should take a leadership role in solid waste management, as indicated by the legislative history of section 211 of the Resource Recovery Act of 1970:

> Federal agencies are inclined to place important environmental quality control functions in a subordinate role to mission. This is no longer appropriate or acceptable. Federal agencies which generate volumes of waste have a correlative responsibility to request appropriations from Congress necessary to properly manage such waste as part of their normal operating expenses. The public will not tolerate the excuse that budget restrictions prevent compliance with waste management standards and guidelines; it is abundantly clear that the provisions of the environmental control laws do not permit the same excuse to be advanced by individuals or private organizations. Federal agencies must take the lead in overcoming the reluctance to invest funds necessary to control solid waste pollution. Senate Report No. 91–1034, Senate Committee on Public Works, 91st Congress, 2nd Session (1970) at p. 14.

Compatibility of Guidelines. There is some overlap in the coverage between these guidelines for resource recovery facilities, the "Source Separation for Materials Recovery Guidelines" and the "Beverage Container Guidelines." Numerous comments were received on the issues of reducing waste and separating waste at the source. Some wanted EPA to require that only waste reduction and source separation measures be taken in the field of solid waste management, while others wanted all waste to be disposed of in resource recovery facilities in lieu of waste reduction and source separation measures.

It is the intent of EPA that agencies should carry out waste reduction, as outlined in the "Beverage Container Guidelines," source separation, as outlined in the "Source Separation for Materials Recovery Guidelines" and recovery of energy and materials as outlined in these guidelines, to the maximum extent possible. After waste reduction and source separation of paper, the remaining wastes should be processed when possible in accordance with these guidelines. Implementation of all three guidelines will result in maximum conservation benefit to the country and economic savings to the government. Additionally, it is desirable and is the clear intent of Congress that the Federal government take a leadership role in the demonstration of techniques for the reduction of waste, the separation of materials at the source, and the utilization of resource recovery facilities.

There may be circumstances where the source separation and recycling of high-grade paper, corrugated containers, or newspaper is economically impracticable due to inability to sell the recovered materials or due to unreasonably high costs. Under such circumstances agencies may choose to recover these materials in centralized recovery facilities or through conversion into energy. The rationale and analysis supporting a decision to choose this form of recovery instead of source separation must be reported to the Administrator under these guidelines.

Technology. There are several systems for recovering materials and energy from solid waste, and these are in various stages of development. The Environmental Protection Agency is participating in the full-scale demonstration of several resource recovery systems under the authority of Section 208(a) of the Solid Waste Disposal Act, as amended. Many communities are designing and building systems patterned after these demonstrations. There is a significant amount of research, development, and system implementation in this field on the part of private industry, universities, States, municipalities, and the Federal Government. (For detailed discussions of the technology involved, see the recommended bibliography at the end of these guidelines).

These guidelines do not recommend any specific technology or system for the recovery of materials or energy because no one system is best under all circumstances. Many different approaches to recovering the energy and materials value from solid waste are presently being examined in a wide variety of communities throughout the country. Not until the investigations recommended in these guidelines have been completed in a thorough manner can a community, agency, or other appropriate entity, considering its disposal and market situation, decide which method will be of greatest benefit. EPA has published a Resource Recovery Implementation Guide that will assist agencies in determining the feasibility of implementing resource recovery, in making the system selection decision, in procuring the selected system, and in marketing the output products.

Determination of Economic Practicability. The legislative history of the Solid Waste Disposal Act, as amended, indicates that it was the intent of Congress that Federal agencies should comply with resource recovery guidelines even when guidelines implementation costs are more than previous solid waste management practices. (Senate Report No. 91–1034, Senate Committee on Public Works, 91st Congress, 2nd Session (1970) at p. 14). However, neither the law nor the legislative history indicates how much additional cost is tolerable. It is not the intent of EPA that resource recovery systems be built in unreasonably uneconomical situations. Therefore, the guidelines state that a valid reason for not implementing a resource recovery system would be that costs of implementation are so high as to be economically impracticable. The base line for determining comparative cost is the cost of complying with the provisions of the "Thermal Processing and Land Disposal of Solid Waste Guidelines" (40 CFR 240 and 241) or the price paid for comparable service under contract. An agency making the judgment whether or not to proceed with a resource recovery system must consider both the base line cost and the Congressional intent.

The precedent for incurring additional costs in order to gain environmental benefits has been clearly established. Section 15(c)(1) of the Noise Control Act (Pub. L. 92–574) requires the procurement of certified low-noise-emission products in lieu of other products if the costs are no more than 125 percent higher. Sections 212G(e) (1) and (2) of the Clean Air Act (Pub. L. 91–604) require the procurement of certified low-emission vehicles in lieu of others if their cost is no more than 150 percent higher and authorize a premium of 200 percent in special instances. The proposed guidelines established a "benchmark" of twice the cost of present environmentally approved solid waste disposal methods as a reasonable cost for resource recovery. After consideration of comments to the effect that the two times cost benchmark could be counterproductive in cases where present disposal costs are low and could have grave economic impact where disposal costs are very high, EPA decided to eliminate any specific reference to cost and to allow each agency to make the economic determination on a case by case basis. Section 245.200–1(f) was changed to reflect this approach.

Facility Size. Several comments were received questioning the minimum size of the resource recovery facilities. These guidelines require all Federal agencies that have jurisdiction over facilities, the operation of which involves disposal of 100 tons or more of solid waste per day after complying with the "Beverage Container" and "Source Separation" guidelines to establish or utilize resource recovery facilities. It should be noted that this tonnage figure is intended to indicate the level above which the establishment or utilization of a resource recovery facility must be investigated. It is neither the intent of these guidelines to require Federal agencies to build resource recovery facilities in unreasonably uneconomical situations, nor to eliminate resource recovery activities at small installations. It is the intent of the guidelines that a resource recovery facility be established or utilized by those Federal facilities which are generating sufficient volumes of solid waste to make resource recovery practicable.

Integration with State and Local Plans. One commenter was concerned with how the establishment of a Federal Resource Recovery Facility would affect local resource recovery efforts. Section 245.200–1(c) was changed in light of this comment to require that resource recovery facilities established as a result of these guidelines be compatible with State and local solid waste management plans.

Processing Percentage. Several comments were received concerning the requirement that established the processing percentage. The commenters assumed, correctly, that a 65 percent resource recovery potential indicated that some specific type of energy recovery technology would be used. One comment pointed out that there was a possibility that energy recovery might not be environmentally sound in certain areas and, therefore, that the 65 percent capability would not be met. Another comment indicated that as the guideline was written, the 65 percent was an all or nothing situation. Obviously, if establishing a resource recovery facility that will process 65 percent of the incoming waste into a marketable product is constrained by other environmental regulations and it is economically impracticable to comply with all environmental regulations, then the provisions of §245.100(g) would apply. Additionally, in order to provide more flexibility in the 65 percent requirement, §245.200−1(e) has been changed to allow the establishment of resource recovery facilities that will process less than 65 percent of the incoming waste into a marketable product when the 65 percent criterion cannot be met because of high cost or the inability to market the output products.

Lead Agency and Responsibility. Because of the difficulty in determining which agency within an SMSA should assume the role of lead agency, EPA has changed the procedures to be used in establishing the lead. The Administrator will analyze agency disposal tonnages and other factors and establish the lead agency in each SMSA. Sections 245.100(g) and (h), and 245.200−1(b) have been changed to reflect this.

The responsibility for implementation of these guidelines has been clarified by changing some wording to match the wording in the Solid Waste Disposal Act of 1965 as amended (Pub. L. 89−272). Sections 245.100(d), 245.101(b), 245.200−1(a), 245.200−1(b), 245.200−2(a) and 245.200−2(b) were changed to reflect the wording of the Act.

Action
These guidelines are issued under the Authority of Section 209(a) of the Solid Waste Disposal Act of 1965 (Pub. L. 89−272) as amended by the Resource Recovery Act of 1970 (Pub. L. 91−512).

Chapter I of Title 40 of the Code of Federal Regulations is amended by adding a new Part 245.

Russell E. Train,
Administrator.

September 10, 1976.

Subpart A—General Provisions
245.100   Scope.
245.101   Definition.

Subpart B—Requirements and Recommended
Procedures
245.200   Establishment or Utilization of Resource Recovery Facilities.
245.200−1   Requirements.
245.200−2   Recommended procedures: Regionalization.
245.200−3   Recommended procedures: Planning Techniques.

Subpart A—General Provisions

§245.100 Scope
(a) These guidelines are applicable to the recovery of resources from residential, commercial, or institutional solid wastes.

(b) The "Requirement" sections contained herein delineate minimum actions for Federal agencies for planning and establishing resource recovery facilities. Pursuant to section 211 of the Solid Waste Disposal Act, as amended, and Executive Order 11752, the "Requirement" sections of this guideline are mandatory for Federal agencies. In addition, they are recommended to State, interstate, regional, and local governments for use in their activities.

(c) The "Recommended Procedures" sections are presented to suggest additional actions or preferred methods by which the objectives of the requirements can be realized. The "Recommended Procedures" are not mandatory for Federal agencies.

(d) These guidelines apply to all Federal agencies that have jurisdiction over any real property or facility the operation or administration of which involves such agency in residential, commercial or institutional solid wastes disposal activities either in-house or by contract. Federal land that is used solely for the disposal of non-Federal solid waste is not considered real property or a facility for the purpose of these guidelines.

(e) The Environment Protection Agency will render technical assistance and other guidance to Federal agencies when requested to do so pursuant to section 3(d) 1 of Executive Order 11752.

(f) Within one year after the final promulgation of these guidelines, agencies shall make a determination as to what actions will be taken to establish a resource recovery facility in accordance with these guidelines and shall, within 60 days of such determination, submit to the Administrator a schedule of such actions.

(g) In order for the Administrator to establish the lead agency in each Standard Metropolitan Statistical Area (SMSA) as addressed in §245.200−1(b), each Agency shall provide the Administrator within 60 days after the final promulgation of these guidelines the following information:

List of all real property or facilities by SMSA that the agency has jurisdiction over, the operation or administration of which involves such agency in residential, commercial or institutional solid wastes disposal activities, either in-house or by contract, in amounts of more than one ton of solid waste per day (equivalent to 260 tons or more annually) after implementation of other Federal guidelines for waste reduction and source separation and that amount of solid waste.

(h) Within 90 days after final promulgation of these guidelines, the Administrator will establish the lead agency in each SMSA.

(i) Federal agencies that make the determination not to establish or utilize a resource recovery facility shall make a report to the Administrator fully explaining that determination. The Administrator shall publish in the Federal Register notice of the availability of this report to the public. In making this determination, agencies must consider energy conservation, environmental factors, and natural resource conservation as well as cost. Trade-offs between these factors must be analyzed prior to the decision not to establish or utilize a resource recovery facility. As all of these factors can be reduced to cost, the following are considered to be valid reasons for not establishing or utilizing a resource recovery facility when supported by individual facts and circumstances:

(1) Costs so high as to render establishing a resource recovery facility economically impracticable; or

(2) Inability to sell the recovered products due to lack of market.

(i) The report required by this section shall contain:

(A) A description of alternative actions considered with emphasis on those alternatives that involve resource recovery, and any actions that would preclude establishing or utilizing a resource recovery facility.

(B) A description of ongoing actions which will be continued and new actions taken or proposed. This statement should identify all agency facilities that will be affected by these actions including a brief description of how these facilities will be affected.

(C) An analysis of the action chosen by the agency including supporting technical data, market studies, and policy considerations so that the factors influencing the decision not to establish a resource recovery facility are clear.

(ii) The report required by this section shall be submitted to the Administrator as soon as possible after a final agency determination has been made not to establish or utilize a resource recovery facility, but in no case later than sixty days after such final determination. The Administrator shall indicate to the agency in writing his concurrence or disagreement with the agency's decision, including his reasons therefor.

(iii) Implementation of actions that would preclude establishing or utilizing a resource recovery facility shall be deferred for 60 days, from the Agency's receipt of the report required by §245.100 (g), in order to give the Administrator an opportunity to receive, analyze and seek clarification of the report.

(iv) It is recommended that where the report required by this section concerns an action for which an Environmental Impact Statement (EIS) is required by the National Environmental Policy Act, that the report be circulated together with the EIS.

§245.101   Definitions

As used in these guidelines:

(a) "Commercial solid waste" means all types of solid waste generated by stores, offices, restaurants, warehouses, and other such non-manufacturing activities, and non-processing waste generated at industrial facilities such as office and packing wastes.

(b) "Disposal" means the collection, storage, treatment, utilization, processing, or final disposal of solid waste.

(c) "Facility" means any building, installation, structure, or public work owned by or leased to the Federal Government. Ships at sea, aircraft in the air, land forces on maneuvers, other mobile facilities, and U.S. Government installations located on foreign soil are not considered "Federal facilities" for the purpose of these guidelines.

(d) "Infectious waste" means: (1) Equipment, instruments, utensils, and fomites (any substance that may harbor or transmit pathogenic organisms) of a disposable nature from the rooms of patients who are suspected to have or have been diagnosed as having a communicable disease and must, therefore, be isolated as required by public health agencies; (2) laboratory wastes, such as pathological specimens (e.g., all tissues, specimens of blood elements, excreta, and secretions obtained from patients or laboratory animals) and disposable fomites attendant thereto; (3) surgical operating room pathologic specimens and disposable fomites attendant thereto and similar disposable materials from outpatient areas and emergency rooms.

(e) "Institutional solid waste" means solid wastes originating from educational, health care, correctional, and other institutional facilities.

(f) "Pyrolytic gas and oil" means gas or liquid products that possess useable heating value that is recovered from the heating of organic material (such as that found in solid waste), usually in an essentially oxygen-free atmosphere.

(g) "Recoverable resources" means materials that still have useful physical, chemical, or biological properties after serving their original purpose and can, therefore, be reused or recycled for the same or other purposes.

(h) "Recovery" means the process of obtaining materials or energy resources from solid waste.

(i) "Recycled material" means a material that is utilized in place of a primary, raw, or virgin material in manufacturing a product.

(j) "Recycling" means the process by which recovered materials are transformed into new products.

(k) "Residential solid waste" means the garbage, rubbish, trash, and other solid waste resulting from the normal activities of households.

(l) "Resource recovery facility" means any physical plant that processes residential, commercial, or institutional solid wastes biologically, chemically, or physically, and recovers useful products, such as shredded fuel, combustible oil or gas, steam, metal, glass, etc. for recycling.

(m) "Tons per day" means annual tonnage divided by 260 days.

Subpart B—Requirements and Recommended Procedures

§245.200 Establishment or Utilization of Resource Recovery Facilities

§245.200-1 Requirements
(a) A Federal agency that has jurisdiction over any real property or facility the operation or administration of which involves such agency in residential, commercial or institutional solid wastes disposal activities either in-house or by contract in amounts of 100 tons or more per day (equivalent to 26,000 tons or more annually) after implementation of other Federal guidelines for waste reduction and source separation shall establish or utilize resource recovery facilities to separate and recover materials or energy or both from such solid waste.

(b) If any one Federal agency within a Standard Metropolitan Statistical Area that has jurisdiction over any real property or facility the operation or administration of which involves such agency in residential, commercial, or institutional solid wastes disposal activities either in-house or by contract in amounts of 50 tons or more per day (equivalent to 13,000 tons or more annually) after implementation of other Federal guidelines for waste reduction and source separation, and if the combined total of these solid wastes for all Federal agencies within the SMSA is 100 tons or more per day (equivalent to 26,000 tons or more annually) after implementation of other Federal guidelines for waste reduction and source separation, all Federal agencies within the SMSA shall establish or utilize one or more resource recovery facilities to separate and recover materials or energy or both from this solid waste. The agency that has jurisdiction over the disposal of the largest quantity of residential, commercial, or institutional solid wastes in the SMSA shall be designated the lead agency by the Administrator of EPA in the resource recovery facility planning process. The lead agency shall be responsible for planning, organizing, and managing the joint resource recovery activities of the agencies in the SMSA and shall report the compliance decision of the agencies in the SMSA in accordance with subparagraph 245.100 (f) or (i), as appropriate, in a consolidated report. All other agencies in the SMSA shall assist in planning such resource recovery activities.

(c)   Agencies shall consult with appropriate State and local agencies, and with concerned local citizens and environmental groups prior to initiation of market analysis and facility design and construction to determine what effects the project might have on local, regional, and State solid waste management plans for the area and to determine the extent of prior resource recovery planning for the area. Resource recovery facilities established as a result of these guidelines shall be compatible with such plans.

(d)   Resource recovery facilities established or utilized as a result of these guidelines shall be designed with a capacity sufficient to process at least all of the residential, commercial, or institutional solid wastes disposed of after implementation of other Federal guidelines for waste reduction and source separation, by the agencies that have jurisdiction over the Federal facilities that will utilize the resource recovery facility.

(e)   Resource recovery facilities established or utilized as a result of these guidelines shall be designed to process at least 65 percent by wet weight of the input solid waste into recycled material, fuel, or energy. Thus, the weight of the unmarketable residue shall be no more than 35 percent by wet weight of the input solid waste. If inability to meet the 65 percent criteria is based on circumstances as stated in §245.100(i) then the processing percentage shall be as great as practicable within those circumstances.

(f)   An agency may determine, under §245.100(i) not to establish or utilize a resource recovery facility when after appropriate analysis it is determined that markets for recovered products are not available, or that the cost of the resource recovery system would be so high as to be economically impracticable.

(g)   Agencies that make the determination not to establish or utilize a resource recovery facility must conduct the analysis required by §245.100(i) at least every three years and report the decision resulting from this analysis to the Administrator in accordance with Section 245.100 (f) or (i), as appropriate.

(h)   In order that the Administrator may fulfill his responsibilities as set forth in EO 11752, section 3 (d,6) to "maintain a continuing review of the implementation of this order," each agency shall, on a yearly basis, submit to the Administrator a report outlining the actions taken by that agency pursuant to these guidelines.

§245.200–2 Recommended Procedures:
Regionalization

(a) Federal agencies that have jurisdiction over facilities within a geographical area should enter into joint resource recovery ventures among themselves and with nearby communities in order to maximize economies of scale.

(b) If a community near a Federal facility operates or is planning to construct a resource recovery facility, the Federal agency having jurisdiction over that facility should participate as appropriate relative to waste load in the financing, construction, and operation of that facility.

§245.200–3 Recommended Procedures:
Planning Techniques

Planning for the implementation of a resource recovery facility should be performed in a systematic manner. A series of reports have been prepared by the Agency's Office of Solid Waste Management Programs. The series, titled Resource Recovery Plant Implementation; Guides for Municipal Officials, should be used as an aid in the planning phase.

(a) Planning and Overview (SW–157.1) provides a framework for the overall planning phase.

(b) Preceding the selection of a specific resource recovery technology, an investigation of markets should be made. Markets (SE–157.3) lists the markets for the recovered materials and outlines steps to be taken to secure those markets.

(c) The various resource recovery methods are covered in Technologies (SW–157.2).

(d) The economic viability of a specific resource recovery facility should be determined only after all costs are accounted for as outlined in Accounting Format (SW–157.6).

(e) Other reports in this series are:

Financing SW–157.4
Procurement SW–157.5
Risks and Contracts SW–157.7
Further Assistance SW–157.8

These reports may be obtained from:
Solid Waste Information Materials Control Section, U.S. Environmental Protection Agency, Cincinnati, Ohio 45268.

[FR Doc. 76–27542 Filed 9–20–76; 8:45 am]

Source: 41 Fed. Reg. 41208–11 (September 21, 1976).

Conclusion

Having already treated the reader to substantial self-descriptive material, this book will not seek further to summarize what has gone before. Rather, a brief perusal of what has not been included will occupy the closing lines. This material may suggest lines of further inquiry in areas analogous to what has been presented.

The purchasing chapters discuss primarily *how* to buy, and, with the exception of Chapter 2's reference to purchasing recycled material, has paid little heed to *what* to buy. In many areas of government operations, labor-intensive machines and processes could be substituted for energy-intensive ones—bicycles might substitute for automobiles, for example. There are doubtless many barriers to such shifts, not the least of which would be employee reluctance, but energy savings might be buttressed by increased employment opportunities.

In addition, the purchasing section touches only tangentially upon the procurement of professional services. Procurement of architectural and engineering services may have substantial impact upon the energy efficiency of government buildings. The American Institute of Architects has done some research on this point [1]. Also, the American Bar Association's Draft Model Procurement Code, Section 4, suggests statutory provisions that would govern this vital area of law.

Another interesting problem involves the financing of retrofitting government buildings (see Chapter 7) at the state and particularly the local level. Modifications to existing structures may not be capable of characterization as "capital" improvements, and this stands as an impediment to the issuing of bonds to cover the cost of modification. Perhaps new types of bonds need to be developed. Certainly, in light of the tremendous energy-saving potential of retrofitting, existing impediments should be removed—and replaced with mechanisms that encourage this strategy.

Finally, an in-depth examination of the potential of government purchasing and operations to foster social goals should be performed. This book has advanced many arguments for the use of these functions to encourage energy conservation. In many ways, these arguments may extend to include other types of social goals—environmental quality, product safety, occupational safety and health, and others. Government operations and procurement may provide an innovative alternative to government regulation in many instances. On the other hand, cluttering the procurement process with too many conditions, exceptions, and other "red tape" can thwart its primary function of obtaining goods, services, and buildings for the government at reasonable cost. What are the limits?

These are but a few of the myriad of topics to be explored in areas relevant to this book and to this series. There are, for example, a special set of problems associated with bringing energy conservation strategies to sparsely populated rural areas. The potential for reducing American energy consumption is vast—and the stakes are high.

NOTES TO CONCLUSION

1. Contact The American Institute of Architects, 1735 New York Avenue N.W., Washington, D.C. 20006. Telephone: (202) 785-7374.

Index

About the Author

Ivan J. Tether, graduate of Georgetown University Law Center, has been with the Environmental Law Institute since early 1975. In addition to his work on the Energy Conservation Project, he has written for the Environmental Law Reporter and is currently researching the application of effluent charges to the control of air and water pollution. During law school, he worked for the Center for Auto Safety and the Occupational Safety and Health Review Commission, as well as the Experiment in International Living. He received his undergraduate degree in political science from Union College in Schenectady, New York.

About the Author

Date Due
